T0259684

SpringerBriefs in Water Science and Technology

More information about this series at http://www.springer.com/series/11214

Buddhi Wijesiri · An Liu
Prasanna Egodawatta · James McGree
Ashantha Goonetilleke

Decision Making with Uncertainty in Stormwater Pollutant Processes

A Perspective on Urban Stormwater Pollution Mitigation

 Springer

Buddhi Wijesiri
Science and Engineering Faculty
Queensland University
of Technology (QUT)
Brisbane, QLD, Australia

James McGree
Science and Engineering Faculty
Queensland University
of Technology (QUT)
Brisbane, QLD, Australia

An Liu
College of Chemistry and Environmental
Engineering
Shenzhen University
Shenzhen, Guangdong, China

Ashantha Goonetilleke
Science and Engineering Faculty
Queensland University
of Technology (QUT)
Brisbane, QLD, Australia

Prasanna Egodawatta
Science and Engineering Faculty
Queensland University
of Technology (QUT)
Brisbane, QLD, Australia

ISSN 2194-7244   ISSN 2194-7252   (electronic)
SpringerBriefs in Water Science and Technology
ISBN 978-981-13-3506-8   ISBN 978-981-13-3507-5   (eBook)
https://doi.org/10.1007/978-981-13-3507-5

Library of Congress Control Number: 2018962764

This Springer imprint is published by the registered company Springer Nature Singapore Pte Ltd.
The registered company address is: 152 Beach Road, #21-01/04 Gateway East, Singapore 189721, Singapore

Contents

Abstract

Improving urban liveability has become challenging for cities around the world due to growing population and changing climate. In this regard, water pollution is a major issue as poor water quality inevitably puts human and ecosystem health at risk. However, the lack of scientific knowledge on processes, which drive urban water pollution, constrains effective pollution mitigation.

Urbanised environments produce pollutants of anthropogenic and natural origin such as particulate solids and toxic heavy metals. These pollutants build-up largely on impervious surfaces such as roads and are subsequently washed off during storm events and discharged into urban receiving waters. Therefore, mitigation of stormwater pollution is a key requirement for improving urban liveability.

Stormwater pollution mitigation relies on the accurate prediction of stormwater quality based on pollutant process modelling tools. This requires the quantification of uncertainty to assess the degree of reliability of stormwater quality predictions. The uncertainty primarily arises from two main sources: (1) use of simplified mathematical replications of pollutant processes in stormwater quality modelling (results in process modelling uncertainty); (2) variability in pollutant processes (results in inherent process uncertainty). However, while the current practice in stormwater quality modelling barely accounts for modelling uncertainty, it does not account for process uncertainty. This constrains the design of effective stormwater pollution mitigation strategies.

This book presents a research study that created fundamental knowledge for understanding how process uncertainty is generated and a scientifically robust approach to quantifying this uncertainty. The study was based on new theoretical developments in the mathematical replication of complex pollutant processes, which were supported by extensive field experiments.

The research study created new knowledge on the intrinsic variability in pollutant build-up and wash-off processes by identifying the characteristics of underlying process mechanisms based on the behaviour of different-sized particles. Further, the correlation between build-up and wash-off processes was also clearly defined. Then, the uncertainty associated with the processes that heavy metals undergo was investigated as a case study. An innovative outcome of this study was the approach developed to quantitatively assess process uncertainty which enabled mathematically incorporating the characteristics of variability in build-up and wash-off processes into stormwater quality models. Additionally, the approach developed can also quantify process uncertainty as an integral part of stormwater quality predictions using common uncertainty analysis techniques.

Further, this book provides a practical approach on how to utilise the research outcomes for informed decision-making for designing effective stormwater pollution mitigation strategies. This approach, based on the concept of logic models, guides stormwater managers to bring academic, industry and technology expertise together to utilise new scientific knowledge through specific activities in order to make informed decisions.

Chapter 1
Understanding Uncertainty Associated with Stormwater Quality Modelling

Abstract Stormwater quality modelling is the common practice for generating information necessary for decision making in the design of stormwater pollution mitigation measures. However, the reliability of modelling outcomes largely depends on two types of uncertainty, namely, uncertainty inherent to stormwater pollutant processes and process modelling uncertainty. The inherent process uncertainty arises due to the intrinsic variability in stormwater pollutant processes. The modelling uncertainty arises from model structure and parameters, and input data and calibration data. The chapter establishes the context for defining and quantifying the uncertainty inherent to pollutant build-up and wash-off processes by bringing together current scientific knowledge from research literature. The temporal changes in particle size results in different particle behaviour during build-up and wash-off, leading to variations in particle-bound pollutant load and composition. Accordingly, the variation in particle size over time can be used as a basis for accounting for process variability in stormwater quality modelling, and thereby assessing process uncertainty.

Keywords Process uncertainty · Particle size · Pollutant build-up
Pollutant wash-off · Stormwater quality · Stormwater pollutant processes

1.1 Background

Stormwater pollution is a major challenge for urban liveability due to the potential risks to human and ecosystem health (Ma et al. 2016, 2017). As depicted in Fig. 1.1, increasing urban population and anthropogenic activities and the spread of the built environment generate large amounts of pollutants that are transported through stormwater runoff to receiving waters. Therefore, the design of effective stormwater pollution mitigation measures is becoming inevitably necessary in urban water management. Such measures need to be developed in a way that toxic stormwater pollutants, which can cause health impacts including neurological and carcinogenic effects (Bocca et al. 2004; Hamers et al. 2002), are removed from the

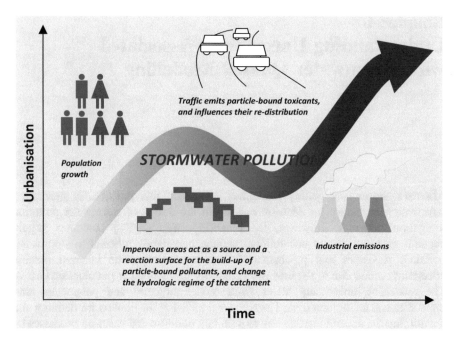

Fig. 1.1 Stormwater pollution in an urban catchment

runoff prior to being discharged into urban receiving waters. This is even more important since stormwater reuse is seen as an alternative water resource, which can supplement depleted natural water resources.

The typical practice of designing measures to mitigate stormwater pollution predominantly relies on stormwater quality modelling that are based on mathematical replications of the primary pollutant processes (Wijesiri et al. 2018). Although there is documented knowledge on the different attributes of these pollutant processes developed over the past decades, the lack of understanding of the uncertainty inherent to pollutant build-up and wash-off constraints the accurate interpretation of stormwater quality modelling outcomes, and consequently, effective decision making in the context of designing pollution mitigation strategies (Barbosa et al. 2012).

1.2   Pollutant Build-up and Wash-off Processes

During dry weather periods, pollutants accumulate on urban surfaces such as roads, and the process is commonly known as build-up. The mechanism of pollutant build-up is driven by the deposition rate, antecedent dry period and re-distribution (Liu et al. 2018). Generally, it is known that pollutants accumulate on the ground surface at a decreasing rate and reach equilibrium between deposition and removal after about 7–9 days (Egodawatta and Goonetilleke 2006). On the other hand, pollutant wash-off occurs during storms as the pollutants accumulated become mobilised and transported through stormwater runoff. The kinetic energy of rain-drops and turbulence of stormwater runoff are the key forces that mobilise and transport the pollutants (Egodawatta et al. 2007).

The mechanisms of pollutant build-up are influenced by several external factors including land-use type, vehicular traffic, street sweeping and wind, while pollutant wash-off is largely dependent on the amount of pollutants available on the ground surface prior to a storm event as well as rainfall characteristics such as rainfall duration and intensity. In fact, most storm events have the potential to wash-off only a fraction of the initially available pollutant load (Egodawatta and Goonetilleke 2006; Egodawatta et al. 2007; Wijesiri et al. 2015).

1.3 Stormwater Quality Modelling

Commercially available stormwater quality modelling tools (e.g. Mike URBAN by Danish Hydraulics Institute and Stormwater Management Model-SWMM by United States Environmental Protection Agency) typically have rainfall-runoff and stormwater quality simulation modules. Given the rainfall characteristics and information on the drainage network, the rainfall-runoff module converts the rainfall into stormwater runoff that is discharged to receiving waters. Primarily based on replication models of pollutant build-up and wash-off processes, the water quality module simulates the pollutants loads generated on urban surfaces and subsequently washed-off by stormwater runoff (Liu et al. 2015).

Generally, stormwater quality modelling approaches can be identified as either deterministic or stochastic. Deterministic approaches generate the same outcomes for a given set of input parameters, while stochastic approaches would produce outcomes with some variability resulting from the probability distributions assigned for each parameter (Obropta and Kardos 2007). Additionally, the capabilities of stormwater quality modelling tools can vary in several other aspects such as the types of pollutants that can be accounted for and the availability of tools for uncertainty assessment.

With regard to the commonly used deterministic and stochastic stormwater quality modelling approaches, most of them only account for the build-up and wash-off of particulate solids, assuming that other pollutants that are attached to solids would have similar behaviour. Thus, these modelling approaches would often ignore potential interactions between solids and other pollutants that could influence stormwater quality in urban catchments. While these approaches also do not undertake accurate process replications (e.g. re-distribution of particulate matter during dry and wet weather periods is not accurately replicated), the lack of uncertainty quantification tools can be identified as a significant drawback (Bicknell et al. 1997; Chen and Adams 2007; Kanso et al. 2003; MikeUrban 2014a, b; Rossman 2009; Wong et al. 2006).

1.4   Uncertainty in Stormwater Quality Predictions

1.4.1   Modelling Uncertainty

Uncertainty in the modelling of pollutant build-up and wash-off processes arises primarily from model input data, model calibration data and model structure. These uncertainty sources, in fact, indicate the lack of completeness of knowledge of the processes being modelled. Therefore, modelling uncertainty is typically referred to as epistemic uncertainty (Dotto et al. 2012; Giudice et al. 2013; Sun et al. 2012).

Epistemic uncertainty is reducible either by bridging the knowledge gaps, or assessed using appropriate uncertainty assessment tools. Quantitative assessment of uncertainty is necessary for the accurate interpretation of model prediction results, and thereby for informed decision making. Typical uncertainty assessment techniques primarily include Classical Bayesian Approach based on Markov Chain Monte Carlo (MCMC) method and the Metropolis-Hastings Sampler (Beven 2009), Multi-algorithm Genetically Adaptive Multi-objective method (AMALGAM) (Vrugt and Robinson 2007), Shuffled Complex Evolution Metropolis Algorithm (SCEM-UM) (Vrugt et al. 2003), and the most widely applied Generalized Likelihood Uncertainty Estimation (GLUE) (Beven and Binley 1992). However, these techniques have limitations in accounting for multiple sources of uncertainty and accurately quantifying uncertainty. The limitations are primarily due to the fact that the assessment of uncertainty relies on user defined likelihood measures (e.g. GLUE method) and prior knowledge (e.g. Bayesian methods). Therefore, the use of more evidence-based approaches for uncertainty assessment is necessary, and the development and application of such an approach are discussed in following chapters.

1.4.2   Process Uncertainty

The changes in the load and composition of pollutants at a particular time during build-up and wash-off create process variability. This variability gives rise to an inherent uncertainty in these processes, which is categorised as aleatory uncertainty (Kiureghian and Ditlevsen 2009). Helton and Burmaster (1996) noted that the lack of knowledge on aleatory uncertainty relates it to epistemic uncertainty (i.e. modelling uncertainty). Accordingly, inaccurate replication of the variability in build-up and wash-off processes in stormwater quality modelling can result in uncertainty in the model predictions. This highlights the importance of in-depth understanding of process variability.

1.5 Variability in Pollutant Processes

As evident from decades of research, in urbanised areas, bulk amounts of particulate solids are found on impervious surfaces (e.g. roads). They are primarily sourced from soil erosion, automobile use and vegetation (Mummullage 2015; Sartor and Boyd 1972). As particulate matter is subjected to external forces driven by both, natural (such as wind) and anthropogenic (such as vehicular traffic) factors, they exhibit different behaviour driven by different physical and chemical characteristics (Gbeddy et al. 2018; Li et al. 2018). As such, particles of different sizes and densities would behave differently during build-up and wash-off, creating process variability (Gunawardana et al. 2013; Wijesiri et al. 2015).

Moreover, the load and composition of pollutants such as metals and hydrocarbons also change during build-up and wash-off, generating process variability. As these pollutants are strongly bound to particles, the behaviour of particles predominantly influence the pollutant load and composition. The role of particle size in creating process variability is further inevitable as different sized particles exhibit different pollutant adsorption behaviour. In fact, particle size brings together several other characteristics of particles, which influence pollutant adsorption behaviour. The characteristics related to particle size include, specific surface area, surface charge density and surface coatings (Jayarathne et al. 2018; Zafra et al. 2011).

1.5.1 Dynamic Behaviour of Particles

Once released into the environment, small and light particles remain in the atmosphere due to slower settling velocities, while large and dense particles undergo direct deposition on urban surfaces during dry weather periods. Particles deposited on surfaces such as roads are continuously subject to re-distribution due to anthropogenic activities (e.g. traffic and street sweeping) and wind. During re-distribution, particle characteristics (primarily physical characteristics) such as particle size are likely to change (Gunawardena et al. 2013; Kupiainen 2007).

As depicted in Fig. 1.2, during re-distribution, particles adhering to impervious surfaces can be detached and re-suspended in the atmosphere when the threshold surface stress is exceeded (Sehmel 1973). This phenomenon commonly occurs due to traffic induced forces. As a result of vehicle movement, turbulent air streams are created above a laminar air stream on road surfaces. As such, particles smaller than the thickness of laminar air flow would remain on the surface, while larger particles are likely to be re-suspended due to the effects of turbulent eddies (Hinds 2012; Mahbub et al. 2011a). The re-suspended particles then interact with particles in the atmosphere and aggregate into even larger particles. The aggregated particles could re-deposit due to increased weight, and then subject to fragmentation by abrasive forces (e.g. tyre abrasion) (Li et al. 2005).

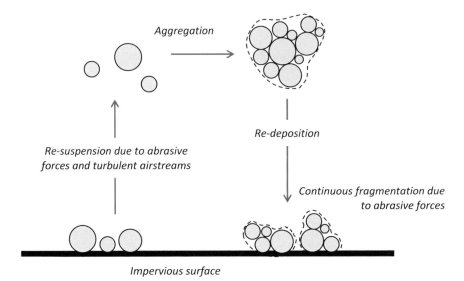

Fig. 1.2 Particle re-distribution during build-up process

Similar to build-up, particles could undergo re-suspension during wash-off due to the impact of raindrops and turbulence created by stormwater runoff. Particles entrained in runoff could aggregate and subsequently re-deposit, and could also be subject to fragmentation. These processes during build-up and wash-off, in fact, significantly changes particle size. Consequently, particle behaviour (e.g. capacity for adsorbing other pollutants) is changed, leading to variations in the load and composition of pollutants attached to particles (Wijesiri et al. 2015).

1.5.2 Pollutant-Particulate Relationships

It is well understood that stormwater pollutants are typically concentrated in the finer particle size fraction. Although the distinction between finer and coarser fraction is subjective, several past studies have shown significantly high concentrations of metals in the particle size fraction <63 µm (Birch and Scollen 2003), <75 µm (Duong and Lee 2009), <125 µm (Manno et al. 2006), <150 µm (Herngren et al. 2006) and <250 µm (Lau and Stenstrom 2005). Similarly, hydrocarbons such as Polycyclic Aromatic Hydrocarbons (PAHs) are found to be concentrated in the particle size fraction < 2 µm (Zhou et al. 2005), <100 µm (Lau and Stenstrom 2005), <150 µm (Herngren 2005), <180 µm (Dong and Lee 2009), <300 µm (Xiang et al. 2010) and <500 µm (Takada et al. 1991).

Moreover, geochemical relationships between pollutants and particles is a major influential factor in the mobility of pollutants during build-up and wash-off, thus resulting in variations in pollutant load and composition. For example, studies based on the sequential extraction of road deposited metals have shown that different species of metals exhibit specific geochemical relationships with different particulate fractions. These include, exchangeable, bound to carbonates, bound to Fe–Mn oxides, bound to organic matter and residual. It has been observed that metals such as Cd and Zn are weakly bound to particles (exchangeable), thus have high mobility, while metals such as Cu and Ni are strongly bound to organic molecular units on the particle surface, thus have low mobility (Banerjee 2003; Duong and Lee 2009).

It is evident that particle size dictates the relationships between particles and other pollutants. Further, it is also important to note that the nature of these relationships can be different between pollutant types, and can potentially change while undergoing build-up and wash-off processes. Such variations in the pollutant affinity to particles are likely to create process variability.

1.5.3 Adsorption Behaviour of Particles

A. Adsorption mechanisms

Particles deposited on urban surfaces adsorb pollutants through the process termed as surface complexation. Pollutants interact with chemically reactive surface functional groups on the particle surface (Sposito 2008). In fact, adsorption is a complex process where pollutants show distinct reactions on the particle surface. As such, pollutants that form the strongest bonds with surface functional groups (ionic and covalent) are attached closest to the particle surface (stern layer in Fig. 1.3). A Stern layer is formed around a particle in response to particle intrinsic charge. Therefore, the particle and the Stern layer have counter charge distributions (Delgado et al. 2007). On the other hand, the pollutants that form electrostatic bonds with surface functional groups are attached relatively loosely to the particle surface (diffuse layer in Fig. 1.3). The diffuse layer is formed due to interactions between positive and negative charges, where the density of charge distribution diminishes away from the particle surface. Further, some pollutants remain electrostatically bound far outside the particle surface, within the diffuse layer (outside slipping plane in Fig. 1.3), but not complexed with surface functional groups (Sparks 2003; Sposito 2008). In fact, the slipping plane is the boundary beyond which the fluid becomes mobile independent of the particle.

Moreover, the electrical charge distribution developed on the particle surface predominantly influences the load and composition of pollutants adsorbed by particles. Charge distribution is significantly influenced by particle physical and chemical characteristics. Therefore, the changes in particle characteristics during build-up and wash-off processes will lead to variations in the load and composition of pollutants discharged to receiving waters.

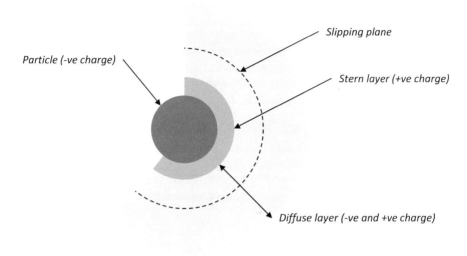

Fig. 1.3 Development of surface charge on a particle surface

B. **Mobility of adsorbed pollutants**

Pollutant mobility varies depending on the nature of the bond between pollutants and particle surface functional groups. As such, pollutants which are complexed with surface functional groups are considered to be immobile with respect to the mobility of particles within the solution. However, the pollutants which are not complexed with surface functional groups are less influenced by particle mobility, in particular, those pollutants existing outside of the slipping plane as shown in Fig. 1.3. This means that the influence of particle mobility on the mobility of adsorbed pollutants decreases with increased distance between the pollutants and the particle surface (Sposito, 2008).

C. **Influential particle characteristics**

The specific surface area of solids (surface area per unit mass/volume) is inversely proportional to the particle size. This means that compared to coarser particles, finer particles have a higher specific surface area, thus have higher surface charge density. Accordingly, pollutants such as metals are more likely to be concentrated in the finer particle size fractions, signifying the influence of the change in particle size on the variability of pollutant load and composition during build-up and wash-off (Cristina et al. 2002; Gunawardana et al. 2012).

Another influential factor in relation to the amount of pollutants adsorbed to a particle surface is metal oxides and organic coatings. Such coatings provide a range of surface functional groups, varying the distribution of positive and negative charges on the particle surface. In this regard, particle mineralogy could also play a key role as it influences the presence of iron, aluminium and manganese. Further, in addition to providing a negative charge that enables the adsorption of cationic pollutants, organic matter also provides surface functional groups that can create organic complexes of pollutants. However, dissolved forms of organic matter, particularly during wash-off, can exert suppression effects on the adsorption of free ions of pollutants such as metals (Gunawardana 2011; Murakami et al. 2008).

1.6   Basis for Assessing Process Uncertainty and Its Implications

It is evident from past literature that: (1) different sized particles behave differently during build-up and wash-off, and in turn create process variability; (2) the strong affinity of pollutants such as metals and hydrocarbons to particulate matter underlines the dependence of pollutants on particle behaviour; and (3) different pollutant adsorption behaviour of different sized particles influence the nature of pollutant affinity to particles, resulting in the variations in pollutant load and composition during build-up and wash-off. Therefore, the temporal change in particle size, in fact, forms a basis for defining the inherent uncertainty in build-up and wash-off due to their intrinsic variability, and thereby providing the basis to quantitatively account for process uncertainty in stormwater quality modelling.

In fact, current uncertainty assessment approaches have limited capacity to account for build-up and wash-off process uncertainty due to poor characterisation of process variability in stormwater quality models. Therefore, it is necessary to develop methodologies to incorporate process variability into replication models of build-up and wash-off processes.

The assessment of process uncertainty should be incorporated into decision making in urban stormwater management. Robust knowledge on stormwater quality is essential in order to formulate the most appropriate stormwater pollution mitigation strategies for a given urban area. Therefore, quantification of uncertainty associated with stormwater quality modelling outcomes would enable accurate interpretation of investigation outcomes, and thereby gain an enhanced understanding of the nature of stormwater pollution in the urban area of interest.

1.7   Summary

Uncertainty associated with the outcomes of stormwater quality models is a challenging issue in the planning and management decision making in relation to urban stormwater pollution mitigation. Two primary types of uncertainties are identified in research literature, namely, uncertainty in the modelling of stormwater pollutant processes and uncertainty inherent to pollutant processes. The modelling uncertainty arises from a number of sources including simplified mathematical replication of pollutant processes and lack of reliable input and calibration data. The inherent process uncertainty is generated from intrinsic process variability. Although modelling uncertainty is generally well understood, process uncertainty needs in-depth understanding, in particular, the characteristics of process variability.

It is important to quantitatively account for the overall uncertainty (modelling and process uncertainty) as an integral part of stormwater quality predictions for the accurate interpretation, and in turn informed decision making. Therefore, understanding how process variability leads to uncertainty and robust methods to incorporate process variability into stormwater quality models are essential for generating more reliable modelling outcomes.

This chapter has synthesised current knowledge on primary stormwater pollutant processes, namely, build-up and wash-off, in order to build a scientifically robust platform for quantitatively accounting for process uncertainty in stormwater quality modelling. Research has confirmed that different sized particles exhibit distinct behaviour during build-up and wash-off. These behaviours can change due to natural and anthropogenic factors common to urban areas. Given that a range of other pollutants such as metals and hydrocarbons are strongly bound to particles, the load and composition of such toxicants are expected to vary in response to particle behaviour. Further, particles of different sizes also exhibit different adsorption behaviour in relation to toxic pollutants, compounding the variability in pollutant load and composition. Moreover, the influence of particle size on the adsorption process predominates compared to other particle characteristics which only play a role in enhancing or constraining adsorption. Accordingly, the change in particle size during build-up and wash-off is the fundamental basis for incorporating process variability into the mathematical replication of build-up and wash-off processes, enabling the quantitative assessment of process uncertainty using appropriate assessment techniques.

References

Banerjee, A. D. K. (2003). Heavy metal levels and solid phase speciation in street dusts of Delhi, India. *Environmental Pollution, 123*(1), 95–105.

Barbosa, A. E., Fernandes, J. N., & David, L. M. (2012). Key issues for sustainable urban stormwater management. *Water Research, 46*(20), 6787–6798.

Beven, K., & Binley, A. (1992). The future of distributed models: Model calibration and uncertainty prediction. *Hydrological Processes, 6*(3), 279–298.

Beven, K. J. (2009). *Environmental modelling: An uncertain future?* London, UK: Routledge.

Bicknell, B. R., Imhoff, J. C., Kittle Jr, J. L., Donigian Jr, A. S., & Johanson, R. C. (1997). Hydrological simulation program-fortran: User's manual version 11. EPA/600/SR-97/080. Athens, GA: U.S. Environmental Protection Agency, National Exposure Research Laboratory.

Birch, G. F., & Scollen, A. (2003). Heavy metals in road dust, gully pots and parkland soils in a highly urbanised sub-catchment of Port Jackson, Australia. *Soil Research, 41*(7), 1329–1342.

Bocca, B., Alimonti, A., Petrucci, F., Violante, N., Sancesario, G., Forte, G., et al. (2004). Quantification of trace elements by sector field inductively coupled plasma mass spectrometry in urine, serum, blood and cerebrospinal fluid of patients with Parkinson's disease. *Spectrochimica Acta, Part B: Atomic Spectroscopy, 59*(4), 559–566. https://doi.org/10.1016/j.sab.2004.02.007.

Chen, J., & Adams, B. J. (2007). A derived probability distribution approach to stormwater quality modeling. *Advances in Water Resources, 30*(1), 80–100.

Chen, M.-H., Fang, H.-W., & Huang, L. (2013). Surface charge distribution and its impact on interactions between sediment particles. *Ocean Dynamics, 63*(9–10), 1113–1121.

Cristina, C., Tramonte, J., & Sansalone, J. (2002). A granulometry-based selection methodology for separation of traffic-generated particles in urban highway snowmelt runoff. *Water, Air, and Soil pollution, 136*(1–4), 33–53. https://doi.org/10.1023/A:1015239831619.

Delgado, A. V., González-Caballero, F., Hunter, R. J., Koopal, L. K., & Lyklema, J. (2007). Measurement and interpretation of electrokinetic phenomena. *Journal of Colloid and Interface Science, 309*(2), 194–224.

Dong, T. T. T., & Lee, B.-K. (2009). Characteristics, toxicity, and source apportionment of polycylic aromatic hydrocarbons (PAHs) in road dust of Ulsan, Korea. *Chemosphere, 74*(9), 1245–1253.

Dotto, C. B. S., Mannina, G., Kleidorfer, M., Vezzaro, L., Henrichs, M., McCarthy, D. T., et al. (2012). Comparison of different uncertainty techniques in urban stormwater quantity and quality modelling. *Water Research, 46*(8), 2545–2558.

Duong, T. T. T., & Lee, B.-K. (2009). Partitioning and mobility behavior of metals in road dusts from national-scale industrial areas in Korea. *Atmospheric Environment, 43*(22–23), 3502–3509.

Egodawatta, P., & Goonetilleke, A. (2006). Chracteristics of pollutants build-up on residential road surfaces. In *Proceedings of the 7th International Conference on HydroScience and Engineering, Philadelphia, USA.*

Egodawatta, P., Thomas, E., & Goonetilleke, A. (2007). Mathematical interpretation of pollutant wash-off from urban road surfaces using simulated rainfall. *Water Res, 41*(13), 3025–3031.

Gbeddy, G., Jayarathne, A., Goonetilleke, A., Ayoko, G. A., & Egodawatta, P. (2018). Variability and uncertainty of particle build-up on urban road surfaces. *Science of the Total Environment, 640–641*, 1432–1437. https://doi.org/10.1016/j.scitotenv.2018.05.384.

Giudice, D. D., Honti, M., Scheidegger, A., Albert, C., Reichert, P., & Rieckermann, J. (2013). Improving uncertainty estimation in urban hydrological modeling by statistically describing bias. *Hydrology and Earth System Sciences, 17*(10), 4209–4225.

Gunawardana, C., Goonetilleke, A., & Egodawatta, P. (2013). Adsorption of heavy metals by road deposited solids. *Water Science and Technology, 67*(11), 2622–2629.

Gunawardana, C., Goonetilleke, A., Egodawatta, P., Dawes, L., & Kokot, S. (2012). Role of solids in heavy metals build-up on urban road surfaces. *Journal of Environmental Engineering, 138*(4), 490–498.

Gunawardana, C. T. K. (2011). *Influence of physical and chemical properties of solids on heavy metal adsorption.* (Ph.D.), Queensland University of Technology.

Gunawardena, J., Egodawatta, P., Ayoko, G. A., & Goonetilleke, A. (2013). Atmospheric deposition as a source of heavy metals in urban stormwater. *Atmospheric Environment, 68*, 235–242.

Hamers, T., Smit, L. A. M., Bosveld, A. T. C., van den Berg, J. H. J., Koeman, J. H., van Schooten, F. J., et al. (2002). Lack of a distinct gradient in biomarker responses in small mammals collected at different distances from a highway. *Archives of Environmental Contamination and Toxicology, 43*(3), 0345–0355. https://doi.org/10.1007/s00244-002-1230-3.

Helton, J. C., & Burmaster, D. E. (1996). Treatment of aleatory and epistemic uncertainty in performance assessments for complex systems. *Reliability Engineering & System Safety, 54*(2–3), 91–258.

Herngren, L., Goonetilleke, A., & Ayoko, G. A. (2006). Analysis of heavy metals in road-deposited sediments. *Analytica Chimica Acta, 571*(2), 270–278.

Herngren, L., Goonetilleke, A., Ayoko, G. A., & Mostert, M. M. M. (2010). Distribution of polycyclic aromatic hydrocarbons in urban stormwater in Queensland. *Australia. Environmental Pollution, 158*(9), 2848–2856.

Herngren, L. F. (2005). *Build-up and wash-off process kinetics of PAHs and heavy metals on paved surfaces using simulated rainfall.* (Ph.D.), Queensland University of Technology.

Hinds, W. C. (2012). *Aerosol technology: Properties, behavior, and measurement of airborne particles.* USA: Wiley.

Jayarathne, A., Egodawatta, P., Ayoko, G. A., & Goonetilleke, A. (2018). Intrinsic and extrinsic factors which influence metal adsorption to road dust. *Science of the Total Environment, 618* (Supplement C), 236–242. doi:https://doi.org/10.1016/j.scitotenv.2017.11.047.

Kanso, A., Gromaire, M.-C., Gaume, E., Tassin, B., & Chebbo, G. (2003). Bayesian approach for the calibration of models: Application to an urban stormwater pollution model. *Water Science and Technology, 47*(4), 77–84.

Kiureghian, A. D., & Ditlevsen, O. (2009). Aleatory or epistemic? Does it matter? *Structural Safety, 31*(2), 105–112.

Kupiainen, K. (2007). *Road dust from pavement wear and traction sanding. Monographs of the Boreal Environment Research*, (Vol. 26). Finnish Environmental Institute, Helsinki, Finland.

Lau, S.-L., & Stenstrom, M. K. (2005). Metals and PAHs adsorbed to street particles. *Water Research, 39*(17), 4083–4092.

Li, Y., Jia, Z., Wijesiri, B., Song, N., & Goonetilleke, A. (2018). Influence of traffic on build-up of polycyclic aromatic hydrocarbons on urban road surfaces: A Bayesian network modelling approach. *Environmental Pollution, 237*, 767–774. https://doi.org/10.1016/j.envpol.2017.10. 125.

Li, Y., Lau, S.-L., Kayhanian, M., & Stenstrom, M. K. (2005). Particle size distribution in highway runoff. *Journal of Environmental Engineering, 131*(9), 1267–1276.

Liu, A., Goonetilleke, A., & Egodawatta, P. (2015). Role of rainfall and catchment characteristics on urban stormwater quality. Singapur: Springer.

Liu, A., Wijesiri, B., Hong, N., Zhu, P., Egodawatta, P., & Goonetilleke, A. (2018). Understanding re-distribution of road deposited particle-bound pollutants using a Bayesian Network (BN) approach. *Journal of Hazardous Materials, 355*, 56–64.

Ma, Y., Liu, A., Egodawatta, P., McGree, J., & Goonetilleke, A. (2016). Quantitative assessment of human health risk posed by polycyclic aromatic hydrocarbons in urban road dust. *Science of the Total Environment.* doi:http://dx.doi.org/10.1016/j.scitotenv.2016.09.148.

Ma, Y., McGree, J., Liu, A., Deilami, K., Egodawatta, P., & Goonetilleke, A. (2017). Catchment scale assessment of risk posed by traffic generated heavy metals and polycyclic aromatic hydrocarbons. *Ecotoxicology and Environmental Safety, 144*(Supplement C), 593–600. doi: https://doi.org/10.1016/j.ecoenv.2017.06.073.

Mahbub, P., Ayoko, G. A., Goonetilleke, A., & Egodawatta, P. (2011a). Analysis of the build-up of semi and non volatile organic compounds on urban roads. *Water Research, 45*(9), 2835–2844.

Mahbub, P., Goonetilleke, A., Egodawatta, P. K., Yigitcanlar, T., & Ayoko, G. A. (2011b). Analysis of build-up of heavy metals and volatile organics on urban roads in Gold Coast, Australia. *Water Science and Technology, 63*(9), 2077–2085.

Manning, M. J., Sullivan, R. H., & Kipp, T. M. (1977). *Nationwide evaluation of combined sewer overflows and urban stormwater discharges Vol. III: Characterization of discharges.* US EPA-600/2-77-064c. Washington, D.C.: U.S. Environmental Protection Agency

Manno, E., Varrica, D., & Dongarrà, G. (2006). Metal distribution in road dust samples collected in an urban area close to a petrochemical plant at Gela, Sicily. *Atmospheric Environment, 40* (30), 5929–5941.

MikeUrban. (2014a). Mike urban collection system—User guide. Danish Hydraulic Institute.

MikeUrban. (2014b). Mouse pollution transport—Reference manual. Danish Hydraulic Institute.

Mummullage, S. W. N. (2015). *Source characterization of urban road surface pollutants for enhanced water quality predictions.* (Ph.D.), Queensland University of Technology (QUT).

Murakami, M., Nakajima, F., & Furumai, H. (2008). The sorption of heavy metal species by sediments in soakaways receiving urban road runoff. *Chemosphere, 70*(11), 2099–2109.

Obropta, C. C., & Kardos, J. S. (2007). Review of urban stormwater quality models: Deterministic, stochastic, and hybrid approaches. *Journal of the American Water Resources Association, 43* (6), 1508–1523.

Rossman, L. A. (2009). Stormwater management model user's manual version 5.0. EPA/600/ R-05/040. Washington, D.C.: U.S. Environmental Protection Agency.

Sartor, J. D., & Boyd, G. B. (1972). *Water pollution aspects of street surface contaminants* EPA-R2-72-081. Washington, D.C.: U.S. Environmental Protection Agency.

Sehmel, G. A. (1973). Particle Resuspension from an asphalt road caused by car and truck traffic. *Atmospheric Environment (1967), 7*(3), 291–309.

Sparks, D. L. (2003). *Environmental soil chemistry*: Access Online via Elsevier.

Sposito, G. (2008). *The chemistry of soils*: Oxford University Press.

Sun, S., Fu, G., Djordjević, S., & Khu, S.-T. (2012). Separating aleatory and epistemic uncertainties: Probabilistic sewer flooding evaluation using probability box. *Journal of Hydrology, 420–421,* 360–372.

Takada, H., Onda, T., Harada, M., & Ogura, N. (1991). Distribution and sources of polycyclic aromatic hydrocarbons (PAHs) in street dust from the Tokyo Metropolitan Area. *Science of the Total Environment, 107,* 45–69.

Vrugt, J. A., Gupta, H. V., Bouten, W. &, Sorooshian, S. (2003). A shuffled complex evolution metropolis algorithm for optimization and uncertainty assessment of hydrologic model parameters. *Water Resources Research, 39*(8).

Vrugt, J. A., & Robinson, B. A. (2007). Improved evolutionary optimization from genetically adaptive multimethod search. *Proceedings of the National Academy of Sciences, 104*(3), 708–711.

Wijesiri, B., Egodawatta, P., McGree, J., & Goonetilleke, A. (2015). Influence of pollutant build-up on variability in wash-off from urban road surfaces. *Science of the Total Environment, 527–528,* 344–350. https://doi.org/10.1016/j.scitotenv.2015.04.093.

Wijesiri, B., Egodawatta, P., McGree, J., & Goonetilleke, A. (2016). Understanding the uncertainty associated with particle-bound pollutant build-up and wash-off: A critical review. *Water Research, 101,* 582–596. https://doi.org/10.1016/j.watres.2016.06.013.

Wijesiri, B., Liu, A., Gunawardana, C., Hong, N., Zhu, P., Guan, Y., et al. (2018). Influence of urbanisation characteristics on the variability of particle-bound heavy metals build-up: A comparative study between China and Australia. *Environmental Pollution, 242,* 1067–1077. https://doi.org/10.1016/j.envpol.2018.07.123.

Wong, T. H. F., Fletcher, T. D., Duncan, H. P., & Jenkins, G. A. (2006). Modelling urban stormwater treatment—A unified approach. *Ecological Engineering, 27*(1), 58–70.

Xiang, L., Li, Y., Yang, Z., & Shi, J. (2010). Influence of traffic conditions on polycyclic aromatic hydrocarbon abundance in street dust. *Journal of Environmental Science and Health Part A, 45* (3), 339–347.

Zafra, C. A., Temprano, J., & Tejero, I. (2011). Distribution of the concentration of heavy metals associated with the sediment particles accumulated on road surfaces. *Environmental Technology, 32*(9), 997–1008.

Zhou, J., Wang, T., Huang, Y., Mao, T., & Zhong, N. (2005). Size distribution of polycyclic aromatic hydrocarbons in urban and suburban sites of Beijing, China. *Chemosphere, 61*(6), 792–799.

Chapter 2
Pollutant Build-up and Wash-off Process Variability

Abstract The outcomes of a series of mathematical simulations of pollutant build-up and wash-off are presented in this chapter, strengthening the knowledge base on process variability. It was found that the build-up of particles <150 μm and >150 μm have different temporal patterns. These patterns could be used to differentiate between the behaviour of particles. The behaviour of particles <150 μm was found to play a key role in creating build-up process variability. On the other hand, the load and composition of different sized particles available on urban surfaces prior to a rainfall event were found to significantly influence the wash-off process. Similar to build-up, wash-off process variability is largely dependent on the behaviour of particles <150 μm, although the contribution from particles >150 μm is significant during rainfall events with relatively shorter duration.

Keywords Particle size · Process variability · Pollutant build-up
Pollutant wash-off · Stormwater quality · Stormwater pollutant processes

2.1 Background

This chapter provides the fundamental knowledge on pollutant build-up and wash-off process variability necessary for developing methodology to assess inherent process uncertainty. Initially, the variations in the behaviour of particles of different size ranges and load and composition of associated pollutants during build-up over the dry weather periods were investigated. Based on the identified characteristics of the build-up process, the variability in the wash-off process was characterised, and the relationship between build-up and wash-off processes was established. The mathematical interpretations of pollutant build-up and wash-off processes presented provide the framework for incorporating process variability into stormwater quality modelling.

This chapter also presents the experimental approach adopted to generate data required for all investigations described in this book. It is important to note that the investigations described in this chapter focus on the build-up and wash-off

© The Author(s), under exclusive license to Springer Nature Singapore Pte Ltd. 2019   15
B. Wijesiri et al., *Decision Making with Uncertainty in Stormwater Pollutant Processes*, SpringerBriefs in Water Science and Technology,
https://doi.org/10.1007/978-981-13-3507-5_2

processes of particulate solids, while subsequent chapters focus on processes in relation to metals as a major particle-bound pollutant found in stormwater runoff.

2.2 Study Approach

The approach adopted for developing methodology to quantitatively assess the inherent process uncertainty was based on establishing a knowledge base in relation to process variability to underpin the mathematical analyses undertaken. In this regard, reliable experimental data played a critical role for accurately formulating the characteristics of pollutant build-up and wash-off. The experimental data used in this study consisted of both, historical and independently collected data. Historical data (see Sect. 2.2.1) was used in the preliminary analyses of the variability in pollutant build-up and wash-off processes (discussed in this chapter) and the quantitative assessment of uncertainty inherent to pollutant build-up and wash-off (discussed in Chap. 3). The new data (see Sect. 2.2.2) collected from field studies was used to validate the outcomes of the historical data analyses and to also further investigate pollutant build-up and wash-off processes (discussed in Chap. 4).

2.2.1 Historical Data

Historical data on particulate build-up and wash-off was obtained from the study conducted by Egodawatta (2007) in the Gold Coast region, South East Queensland, Australia. The road sites (Gumbeel Court, Lauder Court and Piccadilly Place) were located in a residential catchment, namely, Highland Park (Fig. A.1 in Appendix A).

   The build-up data included particulate solids loads for different antecedent dry periods (1, 2, 3, 7, 14, 21 and 23 days) and the wash-off data included particulate solids loads for different rainfall events (20, 40, 65, 86, 115 and 133 mm/h). Further, particle size distributions of 1–900 μm corresponding to each build-up and wash-off sample collected were also obtained. Further details of the characteristics of study sites and sampling and laboratory testing procedures can be found in Egodawatta (2007).

2.2.2 New Data

A. Study catchments

Build-up and wash-off sampling were undertaken at two road sites from two sub-urbs in Gold Coast, South East Queensland, Australia. The study area for this study

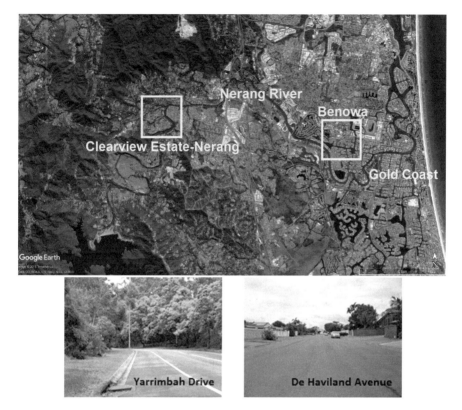

Fig. 2.1 Aerial view of the study catchment and street views of road sites relating to new data

was the same as the area where the historical data was collected. The road sites were located along the largest waterway system in the region, namely, Nerang River, and encompassed a range of different urban forms, traffic conditions, and road surface conditions. This ensured the collection of build-up and wash-off data with significant variability in particle size, which was necessary for the detailed assessment of process variability. The aerial and street views of the road sites are shown in Fig. 2.1 and the details of the characteristics of road sites are given in Table 2.1.

B. **Build-up and wash-off sampling and laboratory analysis**

Build-up sampling

Particulate build-up sampling was undertaken using a dry and wet vacuum system, which has high efficiency in collecting fine particles and minimal generation of additional material (asphalt). Samples were collected from 3 m^2 plots for antecedent dry periods of 2, 4, 5, 7, 8, 10, 12, 19 and 24 days. Each dry period commenced immediately after a rainfall event had occurred in the study

Table 2.1 Characteristics of road sites

Suburb	Road site	Population density[a] (residents/km^2)	Traffic volume and congestion	Road texture[b] depth (mm)
Clearview Estate-Nerang	Yarrimbah Drive	456.6	Moderate	0.84
Benowa	De Haviland Avenue	1173.4	Moderate/low	0.91

[a]ABS (2011)
[b]Adapted from Gunawardana (2011) as measured using the standard sand patch test

area, and all rainfall events were monitored through the records of the Australian Bureau of Meteorology.

Wash-off sampling

Given the practical constraints associated with controlling the intensity and duration of natural rainfall events, wash-off sampling was undertaken using a mechanical rainfall simulator (Herngren 2005). Wash-off samples were collected from road sites, where the maximum build-up was observed during build-up sampling. Average rainfall intensities of 45 and 70 mm/h were simulated over a period of 30 min, and samples were collected in 5-min intervals. These two rainfall intensities were selected for the simulation, as they fall within the range of rainfall intensities used for generating the historical data (i.e. 20–133 mm/h). Additionally, the range of rainfall intensities which the rainfall simulator can reliably simulate was also taken into account. Further details on rainfall simulation can be found in Wijesiri et al. (2015a). At each sampling site, a build-up sample was also collected in order to determine the amount of particulate solids available prior to simulating the rainfall events.

Laboratory testing

The particle size distributions (0.01–3500 μm) of build-up and wash-off samples were analysed using a Malvern Mastersizer 3000 instrument (Malvern Instrument Ltd., 2015). The particulate solids loads in the build-up and wash-off (suspended and dissolved solids) samples were determined for particle size fractions <150 and >150 μm using the test methods 2540B, 2540C and 2540D (APHA 2012). These size ranges were selected as past research has confirmed that the behaviour of particles <150 and >150 μm are the key to creating variability in build-up and wash-off processes (see Sect. 2.3).

For metals (the analysis is discussed in Chap. 4), the size fractionated particulate solids samples were analysed using standard test methods. The analysis involved HNO_3 acid digestion of samples using method 3030E (APHA 2012) and an Environmental Express SC154 hot block digester, in order to extract the metals attached to particulate solids. Subsequently, digested samples were analysed for metals commonly found in the urban environment (Mummullage et al. 2016), namely, Al, Cr, Mn, Fe, Ni, Cu, Zn, Mn and Pb concentrations using Inductively Coupled Plasma-Mass Spectrometry (ICP-MS) using method 200.8 (USEPA 1994) using an Agilent 8800 Triple Quadrupole analyser.

Further, the organic matter content in the build-up and wash-off samples were analysed for both, particulate and dissolved forms. Organic matter is a major influential factor in pollutant adsorption to particulates (as discussed in Chap. 1). The particulate organic fraction was determined using loss-on-ignition method (Rayment and Lyons 2011), and dissolved organic fraction was determined by following the test method 5310C (APHA 2012) using Shimadzu TOC-V$_{CSH}$ instrument.

2.3   Pollutant Processes

2.3.1   Build-up

The analysis of the temporal variations of particulate solids build-up using the historical data showed that particles <150 and >150 µm exhibit significantly different behaviour, forming characteristic build-up patterns (see Fig. A.2 in Appendix A). Over the antecedent dry period, particles <150 µm showed an asymptotically decreasing pattern and particles >150 µm showed an asymptotically increasing pattern. In fact, the two patterns were the inverse of each other and the pattern of particles >150 µm was similar to that of total particulate build-up which is the common representation of the build-up process. Accordingly, it was evident that particles <150 µm are more susceptible to re-distribution and particles >150 µm are likely to undergo continuous deposition. This confirmed that particle size determines particle behaviour, and in turn, build-up process variability.

Accordingly, the behaviours of the two particle size fractions could be mathematically replicated as given in Eq. 2.1 (for particles >150 µm) and Eq. 2.2 (particles <150 µm).

$$B_{(>150)} = a_1 D^{b_1} \qquad\qquad (2.1)$$

$$B_{(<150)} = a_2 D^{-b_2} \qquad\qquad (2.2)$$

where;

B   build-up load
D   antecedent dry period
a1, b1, a2, b2   build-up coefficients.

The basis for the proposed replication models is the power function which was found to be the most consistent with experimental data of total particulate build-up (Ball et al. 1998). Figure 2.2 shows the generalised patterns of particulate build-up.

Fig. 2.2 Generalised depiction of the predicted build-up patterns

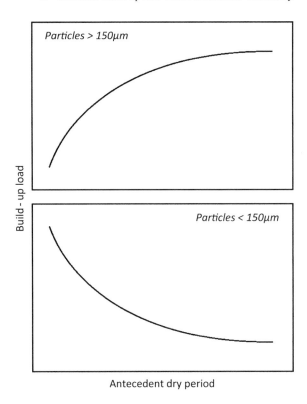

These generalised patterns were derived from the patterns shown in Fig. A.3 in Appendix A, which were predicted by the Eqs. 2.1 and 2.2, using the historical data and non-linear regression. Further details can be found in Wijesiri et al. (2015c). These patterns provide evidence that particles behave differently over the antecedent dry period. Moreover, the particle behaviour would determine the variations in particle-bound pollutant load and composition given that particle size is a key factor in pollutant adsorption.

2.3.2 Wash-off

Unlike build-up, wash-off does not show different patterns for <150 and >150 μm particle size ranges. This is attributed to the limited exposure to re-distribution caused by the duration of storm events (typically in minutes or hours) that is relatively shorter than the antecedent dry period (typically in days). Both particle size fractions showed an increasing pattern, thus wash-off of both particle size fractions could be replicated using the typical exponential decay function as defined by Eq. 2.3 (for particles >150 μm) and Eq. 2.4 (for particles (<150 μm)

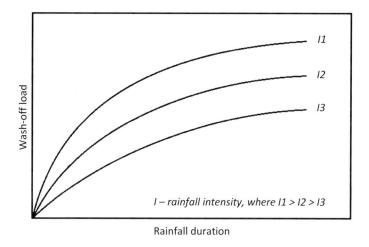

Fig. 2.3 Generalised depiction of the predicted wash-off patterns

(Sartor and Boyd 1972). Figure 2.3 shows the generalised patterns of predicted wash-off for different storm events, which were derived from the patterns predicted by Eqs. 2.3 and 2.4 using non-linear regression and the historical data (Figs. A.4 and A.5 in Appendix A), and further details can be found in Wijesiri et al. (2015b).

$$W_{(>150)} = W_{0(>150)}\left(1 - e^{-k_1 It}\right) \tag{2.3}$$

$$W_{(<150)} = W_{0(<150)}\left(1 - e^{-k_2 It}\right) \tag{2.4}$$

where;

W	wash-off load
W_0	initially available build-up
t	rainfall duration
k1 and k2	wash-off coefficients.

The wash-off coefficient (k) estimated for storm events with different intensities was approximately similar. The average of the k values is given in Table 2.2. The details of the estimation of k value can be found in Wijesiri et al. (2015b). The k

Table 2.2 Estimated wash-off coefficient (k) (adapted from Wijesiri et al. 2015b)

Road site	Wash-off coefficient k (mm^{-1})	
	<150 μm	>150 μm
Gumbeel Court	0.015 (se − 0.001)	0.013 (se − 0.002)
Lauder Court	0.049 (se − 0.003)	0.016 (se − 0.001)
Piccadilly Place	0.036 (se − 0.004)	0.008 (se − 0.0004)

se—standard error

value corresponding to particle size fraction <150 μm is greater than that for particle size fraction >150 μm, implying more rapid wash-off of particles <150 μm compared to particles >150 μm.

2.3.3   Influence of Pollutant Build-up on Variability in Pollutant Wash-off

This investigation involved the mathematical simulation of: (1) wash-off loads of particles size fractions <150 μm and >150 μm; and (2) variations in fraction wash-off—F_w (ratio between wash-off load and initially build-up load) (Egodawatta et al. 2007), using Eqs. 2.3 and 2.4.

It was found that the particulate wash-off varies proportionately with the particulate build-up available prior to a storm event. This in fact confirmed that particulate build-up patterns identified in Sect. 2.3.1 exert a significant impact on the variability in wash-off of different sized particles. Moreover, Fig. 2.4 shows generalised patterns of F_w for particle size fractions <150 μm and >150 μm. The generalised patterns were derived based on the patterns of F_w predicted using non-linear regression and using historical data as shown in Fig. A.6 in Appendix A and further details can be found in Wijesiri et al. (2015b). The differences in the two patterns signify that each particle size fraction influences differently the variation in wash-off load during a storm event. Further, F_w was found to be independent of the initial build-up, but varied over different storm events.

According to Fig. 2.4, F_w of particles <150 μm is not only greater than that of particles >150 μm, but also takes a relatively longer period of time to reach a

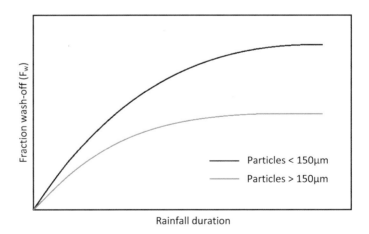

Fig. 2.4 Generalised patterns of fraction wash-off (F_w) (*Note* original predicted patterns can be found in Wijesiri et al. 2015b)

constant value. Thus, compared to particles >150 μm, more rapid variation in the wash-off load of particles <150 μm can be expected during longer-duration storm events. The load and composition of pollutants in the initial build-up of particles <150 μm in turn will determine the load and composition of washed-off pollutants.

2.4 Summary

Particles of different sizes exhibit characteristic behaviours during build-up, such that particles <150 μm exhibits a decreasing pattern and particles >150 μm exhibit an increasing pattern similar to total particulate build-up over the antecedent dry period. These build-up patterns provide insights into the behavioural variability of particles that induce temporal variations in particle-bound pollutant load and composition.

In regard to particulate wash-off, mathematically simulated wash-off load of particle size fractions <150 μm and >150 μm show temporal variations proportional to the initial build-up of each particle size fraction prior to a storm event. Accordingly, the load and composition of particle-bound pollutant wash-off would be a function of the load and composition of the pollutants in the initial build-up of different sized particles. In fact, the variability in pollutant wash-off is significantly influenced by the variability in the build-up of particles <150 μm, while the build-up variability of particles >150 μm plays an important role only in relatively shorter duration storm events.

References

ABS. (2011). Australian Bureau of Statistics *Census-QuickStats*. Retrieved April 15, 2015, from http://www.abs.gov.au/websitedbs/censushome.nsf/home/quickstats?opendocument&navpos=220.

APHA. (2012). *Standard methods for examination of water and wastewater* (E. W. Rice, R. B. Baird, A. D. Eaton L. S. Clesceri Eds. 22 ed.). Washington, D.C.: American Public Health Association, American Water Works Association, Water Environment Federation.

Ball, J. E., Jenks, R., & Aubourg, D. (1998). An assessment of the availability of pollutant constituents on road surfaces. *Science of the Total Environment, 209*(2–3), 243–254.

Egodawatta, P. (2007). *Translation of small-plot scale pollutant build-up and wash-off measurements to urban catchment scale*. (Ph.D.), Queensland University of Technology.

Egodawatta, P., Thomas, E., & Goonetilleke, A. (2007). Mathematical interpretation of pollutant wash-off from urban road surfaces using simulated rainfall. *Water Research, 41*(13), 3025–3031.

Gunawardana, C. T. K. (2011). *Influence of physical and chemical properties of solids on heavy metal adsorption*. (Ph.D.), Queensland University of Technology.

Herngren, L. F. (2005). *Build-up and wash-off process kinetics of pahs and heavy metals on paved surfaces using simulated rainfall*. (Ph.D.), Queensland University of Technology.

Malvern Instrument Ltd. (2015). Mastersizer 3000 User Manual. MAN0474-04-EN-00, UK.

Mummullage, S., Egodawatta, P., Ayoko, G. A., & Goonetilleke, A. (2016). Use of physicochemical signatures to assess the sources of metals in urban road dust. *Science of the Total Environment, 541,* 1303–1309.

Rayment, G. E., & Lyons, D. J. (2011). *Soil chemical methods: Australasia*: CSIRO publishing.

Sartor, J. D., & Boyd, G. B. (1972). *Water pollution aspects of street surface contaminants* EPA-R2-72-081. Washington, D.C.: U.S. Environmental Protection Agency

USEPA. (1994). *Method 200.8: Determination of trace elements in waters and wastes by inductively coupled plasma-mass spectrometry.* Revision 5.4. U.S. Ohio: Environmental Protection Agency.

Wijesiri, B., Egodawatta, P., McGree, J., & Goonetilleke, A. (2015a). Incorporating process variability into stormwater quality modelling. *Science of the Total Environment, 533,* 454–461. https://doi.org/10.1016/j.scitotenv.2015.07.008.

Wijesiri, B., Egodawatta, P., McGree, J., & Goonetilleke, A. (2015b). Influence of pollutant build-up on variability in wash-off from urban road surfaces. *Science of the Total Environment, 527–528,* 344–350. https://doi.org/10.1016/j.scitotenv.2015.04.093.

Wijesiri, B., Egodawatta, P., McGree, J., & Goonetilleke, A. (2015c). Process variability of pollutant build-up on urban road surfaces. *Science of the Total Environment, 518–519,* 434–440. https://doi.org/10.1016/j.scitotenv.2015.03.014.

Chapter 3
Assessment of Build-up and Wash-off Process Uncertainty and Its Influence on Stormwater Quality Modelling

Abstract Current practice in stormwater quality modelling constraints the generation of reliable information about catchment scale stormwater quality due to the lack of robust methods to assess the inherent uncertainty in pollutant build-up and wash-off processes. This chapter presents an approach to quantify process uncertainty as an integral part of stormwater quality predictions. The approach primarily aims to mathematically incorporate the characteristics of process variability into stormwater quality models, and thus quantifying the resulting uncertainty. The application of the new approach revealed that compared to wash-off process uncertainty, the build-up process uncertainty has a greater influence on the prediction of event mean concentrations (EMCs) of particulate solids in urban catchments. Further, it was found that process uncertainty differently influences stormwater quality predictions corresponding to storm events with different intensities, durations and resulting runoff volumes. Planning and management decision making needs to specifically address the changes in the load and composition of particulate solids and associated pollutants during dry weather periods and the storm events that can potentially influence high variations in stormwater quality.

Keywords Particle size · Process uncertainty · Pollutant build-up
Pollutant wash-off · Stormwater pollutant processes · Urban stormwater quality

3.1 Background

The knowledge base on the variability in pollutant build-up and wash-off processes discussed in Chap. 2 was created using small-plot scale field investigations. However, the design of strategies for stormwater pollution mitigation requires knowledge about catchment scale stormwater quality. In this regard, understanding of the uncertainty associated with the predictions of catchment scale stormwater quality is critical for accurate interpretation of those predictions, and in turn, informed planning and management decision making (Liu et al. 2015; Lucke et al. 2018; Ma et al. 2017).

© The Author(s), under exclusive license to Springer Nature Singapore Pte Ltd. 2019   25
B. Wijesiri et al., *Decision Making with Uncertainty in Stormwater Pollutant Processes*, SpringerBriefs in Water Science and Technology,
https://doi.org/10.1007/978-981-13-3507-5_3

Information about stormwater quality in urban catchments is generated using models that describe hydrologic and hydraulic processes within a catchment. As such, the reliability of this information depends on the accuracy of model development, calibration and verification. However, model calibration is a challenging task due to the practical constraints in developing adequate databases. Accordingly, this chapter presents an approach that utilised small-plot scale pollutant build-up and wash-off data (see Sect. 2.2.1 in Chap. 2) to predict catchment scale stormwater quality and to quantify the associated uncertainty. This approach employed MIKE URBAN—SWMM (MikeUrban 2014) software to simulate runoff volumes generated by a number of different natural storm events, and thereby to translate small-plot scale data on pollutant build-up and wash-off into catchment stormwater quality in terms of Event Mean Concentration (EMC). Pollutants simulated were particulate solids since they are the main carrier of other pollutants in stormwater runoff.

The chapter firstly provides details of the study catchments, setting up of catchment models using MIKE URBAN—SWMM, selection of storm events for runoff simulation and the procedure for predicting EMCs for the selected storm events and quantification of uncertainty. Then, the uncertainty associated with the predictions of stormwater quality is characterised with respect to build-up and wash-off processes as well as the characteristics of different storm events.

3.2 Modelling Approach

3.2.1 Catchment Model Setup

Accounting for the spatial variability of hydrologic and hydraulic processes is the key to the accurate setting up of catchment models (Egodawatta 2007). MIKE URBAN—SWMM allows the division of large catchments into small hydrologic units, namely, sub-catchments, based on the layout of the drainage network. The sub-catchments generate stormwater runoff that drains into an adjacent sub-catchment or a node (junction such as a collection point or outfall at the end of a drainage path). Then, pipes and channels (conduits) transport the runoff water from nodes to receiving waters (MikeUrban 2014).

Accordingly, two catchment models were developed for a relatively small catchment—Gumbeel Court (1.6 ha) and a relatively large catchment—Highland Park (105.2 ha) (see Sect. 2.2.1 in Chap. 2). Gumbeel Court model consisted of 2 sub-catchments, 3 junctions, 1 outfall and 3 conduits, while Highland Park model consisted of 625 sub-catchments, 633 junctions, 1 outfall and 635 conduits. The graphical structure of each model is shown in Figs. B.1 and B.2 in Appendix B, and details of the input data used for setting up the catchment models are shown in Figs. B.3–B.8 in Appendix B.

3.2.2 Runoff Simulation

The first step in runoff simulation is to define rainfall records as the primary boundary condition. As such, the daily records of rainfall depth in 1 min interval storm events that have different intensities and durations were obtained from the Australian Bureau of Meteorology (Hinze Dam weather station—040584). From the rainfall records, individual storm events, which produced rainfall depth greater than the depression storage of the catchment impervious area (Wijesiri et al. 2016), were selected for three representative years. The annual rainfall depth over a ten-year period (2004–2014) obtained from the Australian Bureau of Meteorology were analysed to select the representative years as shown in Fig. B.9 in Appendix B. The representative years were considered to reduce the number of storm events in the runoff simulation. Accordingly, individual storm events were selected from two years (27 events from 2005 and 38 events from 2008) where the annual rainfall depth was above and below the average annual rainfall depth (1436.3 mm) and continuous rainfall records were available.

Once the average intensity for each storm event was calculated, the data was input into MIKE Urban—SWMM as a time series to simulate the runoff volume. Table B.1 in Appendix B shows the data in relation to rainfall intensities and durations and simulated runoff volumes.

3.2.3 Stormwater Quality Prediction and Uncertainty Quantification

In this exercise, two scenarios were considered, namely: (1) use of classical build-up and wash-off models (Eqs. 3.1 and 3.2) which do not take into consideration process variability; and (2) use of revised build-up and wash-off models (Eqs. 3.3 and 3.4) incorporated with the characteristics of process variability as discussed in Chap. 2. However, MIKE URBAN—SWMM only allows for in-built models which typically belong to scenario 1. Therefore, MATLAB non-linear regression tools were used to predict stormwater quality and to quantify uncertainty. The procedure for translating small-plot scale stormwater quality into catchment scale and quantification of uncertainty is given in Fig. 3.1.

$$B = at^b \tag{3.1}$$

$$W = (W_o = B)\left(1 - e^{-klt}\right) \tag{3.2}$$

$$B = \left(B_{(<150)} = a_1 t^{-b_1}\right) + \left(B_{(>150)} = a_2 t^{b_2}\right) \tag{3.3}$$

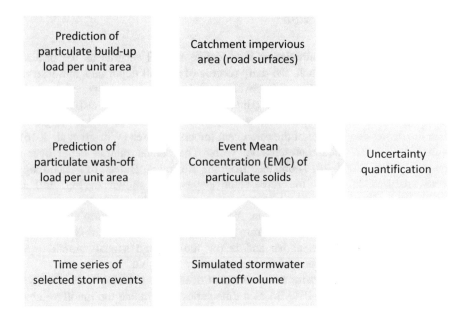

Fig. 3.1 Procedure for translating small-plot scale stormwater quality into catchment scale and quantification of uncertainty

$$W = \left\{ B_{(<150)} \left(1 - e^{-k_1 I t}\right) \right\} + \left\{ B_{(>150)} \left(1 - e^{-k_2 I t}\right) \right\} \tag{3.4}$$

where;

B	build-up load
W	wash-off load
W_o	initially available build-up
t	antecedent dry period/rainfall duration
a, b, a_1, b_1, a_2, b_2, k, k_1 and k_2	build-up and wash-off coefficients.

The first step was to estimate the coefficients for both, build-up and wash-off models given the small-plot scale field data which represent the catchment. It is important to note that the wash-off coefficient (k) should be estimated, such that a common value will represent all the selected storm events. This is due to the similarity in the wash-off coefficient estimated for different storm events (see Sect. 2.3.2 in Chap. 2), and in order to reduce the complexity in the prediction of wash-off loads. As the selected storm events for the years 2005 and 2008 ranged from 13.8 to 74.2 mm/h, the wash-off coefficient was estimated using the data corresponding to storm events simulated for intensities of 20, 40, 65 and 86 mm/h (see Sect. 2.2.1 in Chap. 2). Estimated coefficients of build-up and wash-off models are given in Table B.2 in Appendix B.

Given the estimated coefficients, particulate build-up load per unit area can be predicted using Eqs. 3.1 and 3.3. The predicted build-up load can then be used to predict particulate wash-off load using Eqs. 3.2 and 3.4. These small plot-scale stormwater quality predictions can be translated to catchment scale EMC using the catchment area and the simulated runoff volume for each storm event.

The uncertainty in the predicted stormwater quality was quantified by simulating a significantly large number of EMCs (10,000). This simulation accounted for residual errors and estimation errors for build-up and wash-off coefficients. This was done based on the generic Equations 3.5 and 3.6. Equation 3.5 defines the proportional error model incorporating residual errors (Eq. 3.6) associated with build-up and wash-off models defined by Eqs. 3.1–3.4. Further, estimation errors in build-up and wash-off coefficients were accounted through the covariance matrix of the coefficients of build-up and wash-off models. This enabled the accounting of the uncertainties associated with catchment scale stormwater quality predicted using build-up and wash-off models together with runoff simulated by MIKE URBAN.

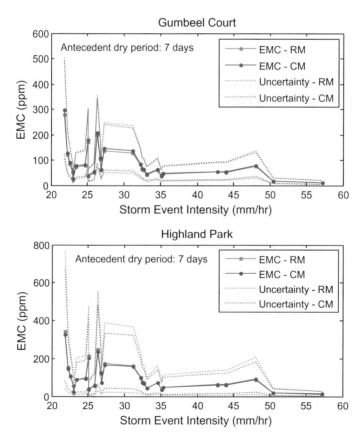

Fig. 3.2 Predicted event mean concentration (EMC) of particulate solids and associated uncertainty—antecedent dry period of 7 days and storm events in 2005. *Note* CM—classical models, and RM—revised models

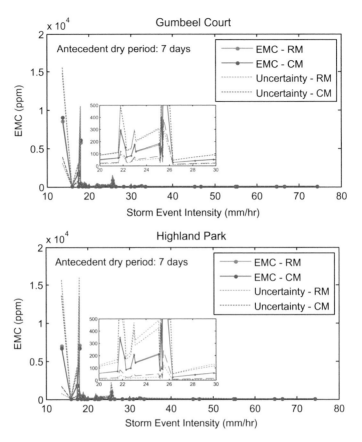

Fig. 3.3 Predicted event mean concentration (EMC) of particulate solids and associated uncertainty—antecedent dry period of 7 days and storm events in 2008. *Note* CM—classical models, and RM—revised models

The uncertainty limits quantified are shown in Figs. 3.2 and 3.3 (7 antecedent dry days) and Figs. B.10–B.13 (1, 2, 3, 14 and 23 antecedent dry days) in Appendix B.

$$y_{sim} = f(1 + \sigma\varepsilon) \qquad (3.5)$$

$$\sigma = \sqrt{\sum_{i=1}^{n} \left(((f - y_{obs})/f)^2 / n \right)} \qquad (3.6)$$

where;

y_{sim} simulated build-up/wash-off
y_{obs} observed (measured) build-up/wash-off

f function value of build-up/wash-off model
ε random scaler drawn from the standard normal distribution
n number of samples.

3.3 Understanding the Influence of Process Uncertainty on Stormwater Quality Predictions

3.3.1 Quantification of Process Uncertainty

The accounting of process uncertainty can be understood by comparing the uncertainty limits quantified for classical and revised models given that classical models are not incorporated with the variability in pollutant build-up and wash-off. As such, Figs. 3.2 and 3.3 and Figs. B.10–B.13 show an increase in the uncertainty bandwidth associated with the EMCs predicted using the revised models (range between the green dotted lines). Further, the difference in the uncertainty bandwidth between classical and revised models was compared using the uncertainty bandwidth normalised by the predicted EMC, which is termed as relative uncertainty bandwidth (RUB). The calculated RUBs are shown in Fig. 3.4 (for 2005) and Fig. B.14 (for 2008) in Appendix B. It is evident that RUB values for the revised models are higher than those for the classical models (compare y-axis between revised and classical models). For example, values of RUB for Gumbeel Court for the revised model is around 1.35 to 1.6, while the corresponding values for classical model are around 1.2–1.3, and a similar pattern can be seen in relation to Highland Park. This confirms the increase in uncertainty quantified using the revised models. Accordingly, these changes in uncertainty, which can be attributed to the changes made to the mathematical form of the classical build-up and wash-off models, indicate that the revised models enable the quantification of process uncertainty.

3.3.2 Differentiation Between Build-up and Wash-off Process Uncertainty

According to Figs. 3.2 and 3.3 and Figs. B.10–B.13, process uncertainty contributes mostly to the change in upper uncertainty limit, implying potentially high variations in predicted stormwater quality resulting from the variability in build-up and wash-off processes. This signifies the importance of accurate quantification of process uncertainty, such that the planning and management decisions made without considering process uncertainty could lead to failure in stormwater pollution mitigation strategies.

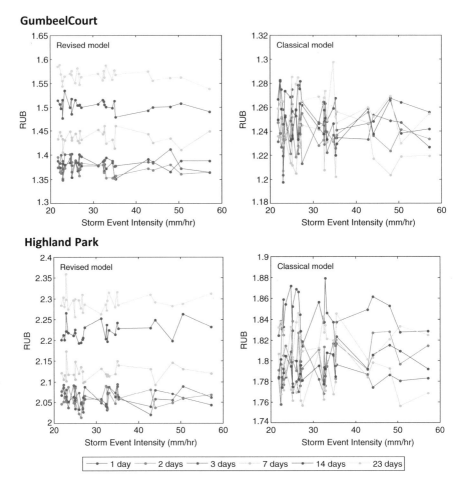

Fig. 3.4 Relative uncertainty bandwidth (RUB) for classical and revised models—storm events in 2005

According to Fig. 3.4 and Fig. B.14, uncertainties in build-up and wash-off processes exert different influences on the predictions of EMCs. To understand this, it is important to note that the change in RUBs between different data series informs the uncertainty in the build-up process, while the change in RUBs along an individual data series informs about the uncertainty in the wash-off process. In relation to the revised models, RUBs of different data series have a wider range than the RUBs for the individual data series. This implies that compared to the uncertainty in wash-off process, the uncertainty in build-up process could contribute to overall uncertainty in stormwater quality predictions, and in turn, the variability in the build-up process as a critical factor that determines the changes in catchment stormwater quality. Therefore, understanding the changes in load and composition of particle-bound toxicants during dry weather periods would be the key to the design of strategies to mitigate urban stormwater pollution.

3.3.3   Characterisation of Process Uncertainty Based on Rainfall Characteristics

Past research studies have shown that storm events that have different intensities and durations could result in significantly varying pollutant loads in the runoff (Liu 2011). Accordingly, the storm events selected for this study were categorised into four types using Intensity-Frequency-Duration (IFD) distributions as shown in Fig. 3.5.

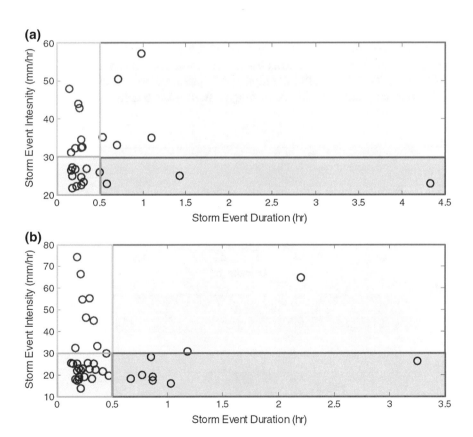

Fig. 3.5 Intensity-frequency-duration (IFD) distribution of storm events; **a** Year 2005; **b** Year 2008; *Note* Highlighted in Green: Type 1 (low intensity-short duration); Red: Type 2 (low intensity-long duration); Yellow: Type 3 (high intensity-short duration); Blue: Type 4 (high intensity-long duration) storm events (adapted from Wijesiri et al. 2016)

- Type 1: low intensity (<30 mm/h) short duration (<0.5 h) events
- Type 2: low intensity (<30 mm/h) long duration (>0.5 h) events
- Type 3: high intensity (>30 mm/h) short duration (<0.5 h) events
- Type 4: high intensity (>30 mm/h) long duration (>0.5 h) events.

It is evident from the IFDs shown in Fig. 3.5 that in both, 2005 and 2008, low intensity-short duration (Type 1) events have occurred most frequently. Although Fig. 3.5 accounts only for rainfall intensity and duration as the major characteristics, the runoff volume could also play a critical role in the effectiveness of stormwater pollution mitigation strategies. For example, it is accepted that the removal of pollutants in smaller volumes of runoff is more effective than treating larger volumes (Guo and Urbonas 1996; Liu 2011). Figures 3.6 and 3.7 show the Intensity-Runoff-Duration distributions for the selected storm events. Type 1 and Type 3 storm events would generate significantly smaller volumes of runoff compared Type 2 and Type 4 storm events.

Moreover, 70% of total storm events belong to Type 1 and Type 3 which have an average intensity of 30 mm/h (see Table B.1). Further, storm events with an average intensity of 30 mm/h and duration <0.5 h would contribute to generating the highest EMCs in the catchments studied (see Figs. 3.2, 3.3 and Figs. B.10–B.13 in Appendix B). Additionally, the high EMCs predicted are associated with relatively wide uncertainty limits. Accordingly, stormwater quality is likely to vary

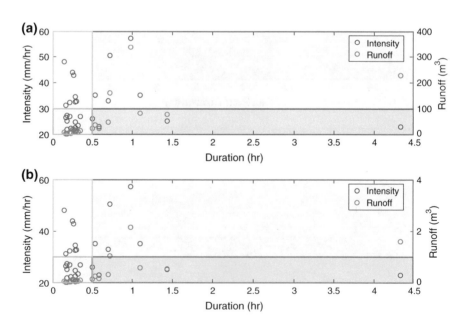

Fig. 3.6 Intensity-runoff-duration distribution of storm events—year 2005; **a** Gumbeel; **b** Highland Park; *Note* Highlighted in Green: Type 1 (low intensity-short duration); Red: Type 2 (low intensity-long duration); Yellow: Type 3 (high intensity-short duration); Blue: Type 4 (high intensity-long duration) storm events

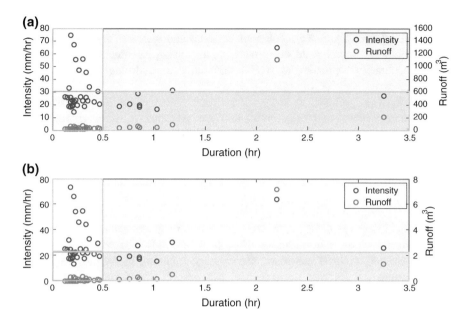

Fig. 3.7 Intensity-runoff-duration distribution of storm events—year 2008; **a** Gumbeel; **b** Highland Park; *Note* Highlighted in Green: Type 1 (low intensity-short duration); Red: Type 2 (low intensity-long duration); Yellow: Type 3 (high intensity-short duration); Blue: Type 4 (high intensity-long duration) storm events

over a wider range during Type 1 and Type 3 storm events compared to Type 2 and Type 4 storm events.

Furthermore, uncertainty limits of EMCs predicted for Type 1 and Type 3 storm events using revised models show variations over a wider range than those EMCs predicted for Type 2 and Type 4 storm events (see <30 mm/h events in Figs. 3.2 and 3.3 and Figs. B.10–B.13). This highlights the importance of in-depth understanding of how inherent process uncertainty could influence stormwater quality under complex environmental conditions.

3.4   Summary

The chapter presents an approach to quantifying the pollutant build-up and wash-off process uncertainty as an integral part of the predictions of catchment stormwater quality. The analysis of stormwater quality modelling outcomes led to creating important knowledge that can be applied to the development of tools for informed decision making in the context of urban stormwater pollution mitigation. The key findings are listed below.

- Given the intensity and duration of typical storm events occurring in an urban catchment, small plot-scale pollutant build-up and wash-off data can be translated into catchment scale storm water quality using catchment models, which can accurately replicate the spatial variability of hydrologic and hydraulic processes.
- Uncertainty in pollutant build-up and wash-off processes can be quantified once the characteristics of process variability are mathematically incorporated into stormwater quality models.
- Uncertainty in the build-up process exerts greater influence on the predictions of stormwater quality compared to the wash-off process. As such, variations in the load and composition of particles and associated pollutants over dry weather periods can drive the changes in stormwater quality in urban catchments.
- In urban catchments in the Gold Coast region, Australia, storm events with an average intensity of 30 mm/h and duration <0.5 h are expected to generate significantly high concentrations of particulate solids. Further, uncertainty in build-up and wash-off processes was found to influence the stormwater quality predictions for different storm events. Accordingly, the design of stormwater pollution mitigation strategies needs to consider not only the characteristics of typical storm events and resulting runoff quantity, but also the uncertainties that influence the reliability of stormwater quality predictions.

References

Egodawatta, P. (2007). *Translation of small-plot scale pollutant build-up and wash-off measurements to urban catchment scale*. (Ph.D.), Queensland University of Technology.

Guo, J. C. Y., & Urbonas, B. (1996). Maximized detention volume determined by runoff capture ratio. *Journal of Water Resources Planning and Management, 122*(1), 33–39. https://doi.org/10.1061/(ASCE)0733-9496(1996)122:1(33).

Liu, A. (2011). *Influence of rainfall and catchment characteristics on urban stormwater quality* (Ph.D.), Queensland University of Technology.

Liu, A., Goonetilleke, A., & Egodawatta, P. (2015). *Role of rainfall and catchment characteristics on urban stormwater quality*. Berlin: Springer.

Lucke, T., Drapper, D., & Hornbuckle, A. (2018). Urban stormwater characterisation and nitrogen composition from lot-scale catchments—New management implications. *Science of the Total Environment, 619*, 65–71.

Ma, Y., McGree, J., Liu, A., Deilami, K., Egodawatta, P., & Goonetilleke, A. (2017). Catchment scale assessment of risk posed by traffic generated heavy metals and polycyclic aromatic hydrocarbons. *Ecotoxicology and Environmental Safety, 144*(Supplement C), 593–600. doi: https://doi.org/10.1016/j.ecoenv.2017.06.073.

MikeUrban. (2014). *Mike urban collection system—User guide*. Danish Hydraulic Institute.

Wijesiri, B., Egodawatta, P., McGree, J., & Goonetilleke, A. (2016). Assessing uncertainty in stormwater quality modelling. *Water Research, 103*, 10–20. https://doi.org/10.1016/j.watres.2016.07.011.

Chapter 4
Case Study—Uncertainty Inherent in Metals Build-up and Wash-off Processes

Abstract This chapter presents the outcomes from a case study which investigated how uncertainty inherent in build-up and wash-off of metals commonly present in urban catchments, could influence stormwater quality. The investigation found consistent and significantly high concentrations of Al, Cr, Mn, Fe, Ni, Cu, Zn, Cd and Pb in the particle size fraction <150 µm than in the particle size fraction >150 µm. When considering consecutive events of build-up and wash-off, the temporal variations in the build-up loads of metals associated with particle size fractions <150 and >150 µm were not consistent with their wash-off loads. These inconsistencies could be potentially due to the interactions between metals and particles that are determined by the particle physico-chemical characteristics. While particle behaviour was found to drive the variability in metal build-up and wash-off, the need for characterising process variability in stormwater quality modelling was highlighted, enabling the quantitative assessment of process uncertainty associated with stormwater quality predictions.

Keywords Metals · Pollutant build-up · Pollutant wash-off · Process uncertainty Process variability · Stormwater quality

4.1 Background

It can be understood from the previous chapters that different behaviours of particles, in particular, particles <150 and >150 µm, create variability in pollutant build-up and wash-off processes, thus resulting in changes in stormwater quality. However, it is not well understood how the load and composition of particle-bound pollutants could change in response to particle behaviour (Wijesiri et al. 2018). This constraints the accurate characterisation of the variability caused by these pollutants using stormwater quality models, and in turn, the assessment of the uncertainty associated with the knowledge generated by those models (Gbeddy et al. 2018).

   The primary objective of this chapter was to lay the foundation to assess the
uncertainty in particle-bound pollutant (metals as a case study) build-up and
wash-off processes. As such, new knowledge on the characteristics of process
variability created through an investigation of eight different species of
particle-bound metals (Zn, Al, Fe, Mn, Cu, Cd, Cr and Pb) is presented in this
chapter. The investigation was undertaken using the data collected through an
independent field study conducted on two road surfaces (Yarrimbah Drive and De
Haviland Avenue) located in Gold Coast region, Australia (see Sect. 2.2.2 in
Chap. 2 for further details). The research outcomes are expected to primarily
contribute to improving the mathematical formulations of particle-bound pollutant
processes in stormwater quality models, and thereby enhancing the accuracy of
uncertainty assessment in stormwater quality predictions. This will assist in
informed planning and management decision making in the context of stormwater
pollution mitigation.

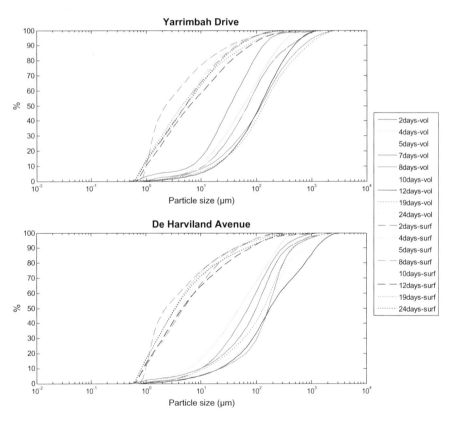

Fig. 4.1 Particle size distributions for build-up. *Note* Volume-based—vol; Surface area-based—
surf

4.2   Influence of Particle Size on Metal Build-up and Wash-off

The surface area-based and volume-based particle size distributions of build-up (Fig. 4.1) and wash-off samples (Fig. 4.2) confirmed that smaller particle sizes have higher particle surface area to volume ratio. This implies that smaller particles could contain relatively large number of surface functional groups that are responsible for the adsorption of pollutants such as metals (see Sect. 1.5.3 in Chap. 1). Accordingly, the majority of metals would be concentrated in the fine particle size fraction.

In fact, the distributions of the concentrations of metals in particle size fractions <150 and >150 μm during build-up (Figs. 4.3 and 4.4) show that particles <150 μm contain the highest concentrations, except for a few inconsistencies around 7 dry days. Moreover, similar observations could be made for the wash-off of most particle-bound metals as evident from Figs. 4.5 and 4.6. This is also supported by the distribution of metal loads between the two particles size fractions as shown in Fig. 4.7. The variability in the metal build-up and wash-off processes,

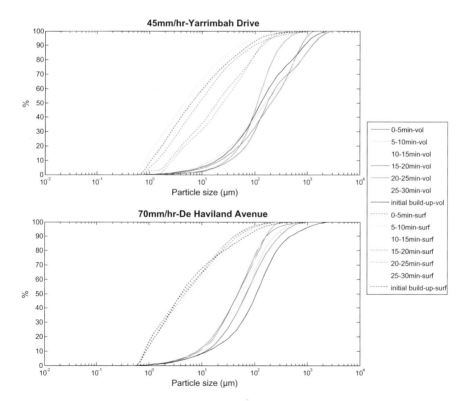

Fig. 4.2 Particle size distributions for wash-off. *Note* Volume-based—vol; Surface area-based—surf

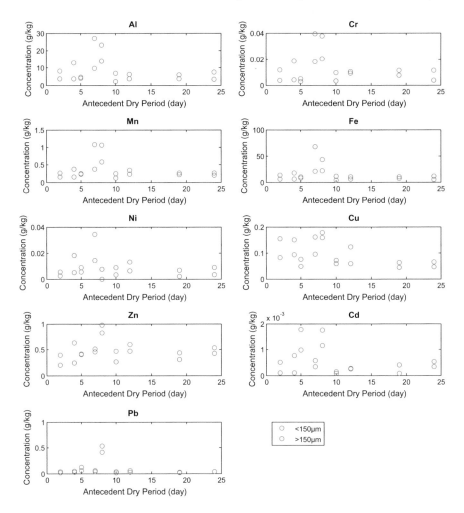

Fig. 4.3 Metals concentrations in particle size fractions <150 and >150 μm during build-up at Yarrimbah Drive

as discussed in the following sections, are based on temporal variations in build-up and wash-off of metals in particle size fractions <150 and >150 μm.

4.3 Variability in Particle-Bound Metal Build-up and Wash-off

Given the strong and consistent affinity of metals to different sized particles (see Sect. 4.2), the patterns of particulate build-up and wash-off described in Sect. 2.3 in Chap. 2 could be effective in understanding the variability in metals build-up and

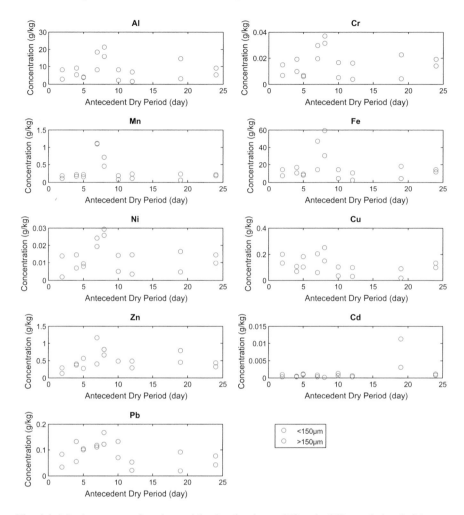

Fig. 4.4 Metals concentrations in particle size fractions <150 and >150 μm during build-up at De Haviland Avenue

wash-off. Accordingly, combinations of metals build-up and wash-off events were considered. In each combination, the wash-off event would precede the build-up event on a continuous timeline. Nine such combinations, which corresponded to the nine antecedent dry periods, could be identified from the build-up and wash-off data for the two road sites (Yarrimbah Drive and De Haviland Avenue).

Figures 4.8 and 4.9 show the combinations of metal build-up and wash-off events. In fact, Figs. 4.8 and 4.9 only depict the build-up events as the corresponding wash-off events were the result from naturally occurring storm events. As such, the build-up load of metals represents not only the metals accumulated during antecedent dry periods, but also the fraction of metals retained after the preceding

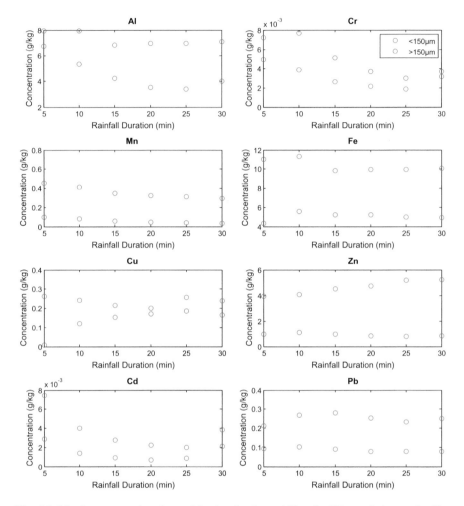

Fig. 4.5 Metals concentrations in particle size fractions <150 and >150 μm during wash-off—45 mm/h event at Yarrimbah Drive

wash-off event. However, it was still possible to observe the patterns of metal wash-off at the two road sites through the simulated storm events. These patterns are shown in Fig. 4.10 in terms of the cumulative load of retained metals after each simulated storm event. This would enable understanding the relationship between temporal variations in metal build-up and wash-off.

According to Figs. 4.8 and 4.9, metal build-up in the particle size fraction <150 μm is greater than that in the particle size fraction >150 μm, which is generally consistent over most antecedent dry periods. Similarly, consistent distributions of most metals in the retained amounts of particle size fractions <150 and >150 μm during wash-off are also present according to Fig. 4.10. Given the

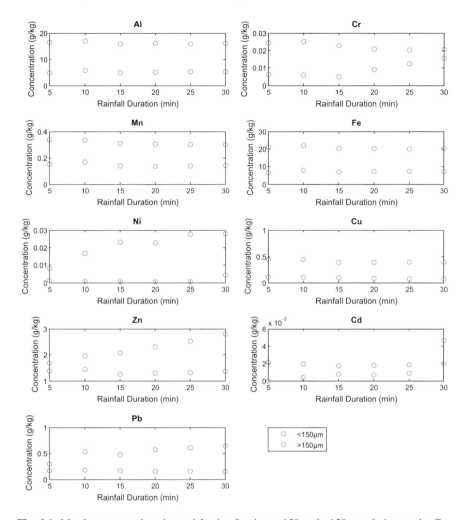

Fig. 4.6 Metals concentrations in particle size fractions <150 and >150 μm during wash-off—70 mm/h event at De Haviland Avenue

strong affinity of metals and the particle size fractions <150 and >150 μm, it is likely that the behaviour of metals during build-up and wash-off would be consistent with that of the two particle size fractions (as discussed in Sect. 2.3 in Chap. 2).

Moreover, it is important to note that there are some metal species that do not show consistent distributions between particle size fractions <150 and >150 μm during build-up and wash-off (Figs. 4.3, 4.4, 4.5 and 4.6). This is attributed to the potential interactions between metals and different sized particles (e.g. adsorption and desorption), implying the significance of the metals adsorption to particulates in

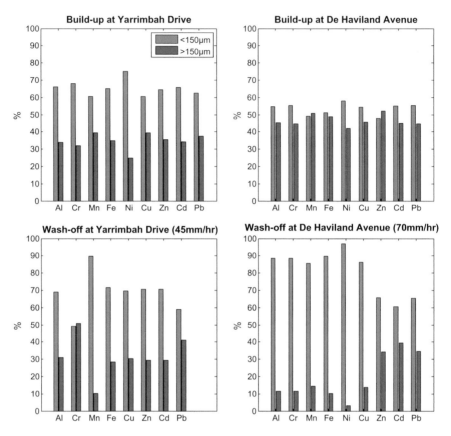

Fig. 4.7 Metals loads (average over antecedent dry period/duration of storm event) in particle size fractions <150 and >150 μm during build-up and wash-off

creating variability in build-up and wash-off processes, and in turn, the stormwater quality.

4.4 Implications of Process Uncertainty on Stormwater Quality Prediction

Changes to stormwater quality at a given point in time can be considered as critical information that is necessary for informed decision making in the mitigation of stormwater pollution. As such, temporal changes in load and composition of particle-bound metals during the build-up and wash-off processes (i.e. process variability) would be essential as metals are one of the abundant toxic pollutants transported by stormwater runoff. In fact, such critical information can be generated

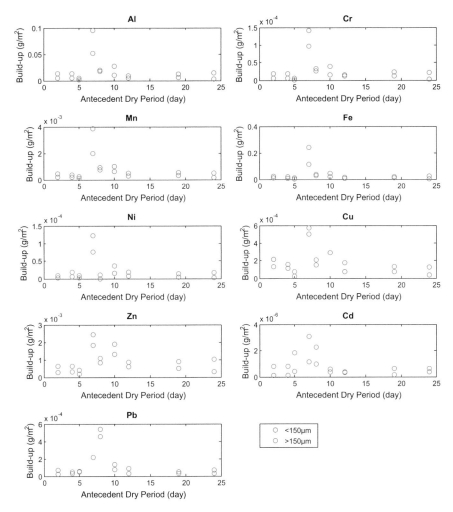

Fig. 4.8 Scenarios of metals build-up and wash-off events at Yarrimbah Drive. *Note* Only build-up event is shown in the figure as the corresponding wash-off events occurred during natural storms (adapted from Wijesiri et al. 2016)

by quantifying the inherent process uncertainty that results from process variability. As noted by Wijesiri et al. (2015), accurate quantification of process uncertainty requires the mathematical formulation of the characteristics of process variability in stormwater quality models. However, current commercially available modelling tools such as MIKE URBAN particularly focus on process replications of particulate solids (MikeUrban 2014). Current, stormwater pollution mitigation strategies also aim to control solids in stormwater runoff. However, these strategies do not address the changes to stormwater quality specifically due to pollutants attached to solids. Given that the outcomes of this study indicate the existence of potentially

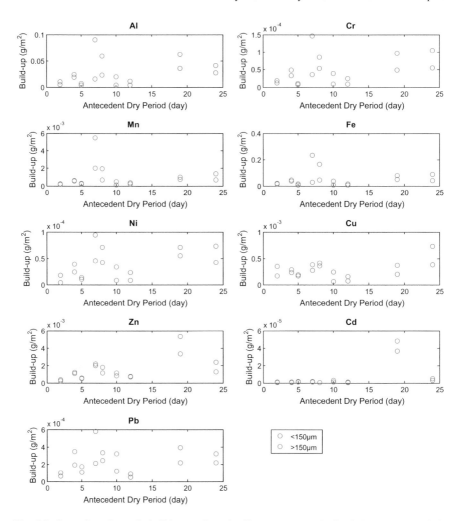

Fig. 4.9 Scenarios of metals build-up and wash-off events at De Haviland Avenue. *Note* Only build-up event is shown in the figure as the corresponding wash-off events occurred during natural storms (adapted from Wijesiri et al. 2016)

complex interactions between metals and particulate solids and the resulting changes to metal load and composition during build-up and wash-off, it is necessary improve model descriptions by incorporating the characteristics of variability in particle-bound pollutant build-up and wash-off. This will enable the generation of reliable information on stormwater quality, and in turn, contribute to formulating effective stormwater pollution mitigation strategies.

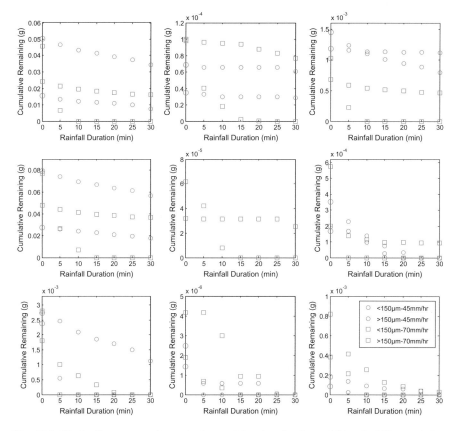

Fig. 4.10 Wash-off patterns of metals in particle size fractions <150 and >150 μm during 45 mm/h event at Yarrimbah Drive and 70 mm/h event at De Haviland Avenue. *Note* initially available load of metals is shown in the vertical axis (adapted from Wijesiri et al. 2016)

4.5   Summary

This chapter provides insights into the variability characteristics of the build-up and wash-off processes of particle-bound metals that would influence assessing the inherent uncertainty of these processes. High concentrations of metals were found in association with the particle size fraction <150 μm and this distribution was also consistent over different field conditions. However, for some species of metals, the typical distribution was found to be different, such that they were concentrated in the particle size fraction >150 μm possibly due to the complex interactions between particulates resulting from adsorption and desorption, particularly during wet weather conditions.

Specific affinity of metals for particle size fractions <150 and >150 μm implied that metals would follow the build-up and wash-off patterns of the corresponding particle size fraction. The case study confirmed that the behavioural variability of different sized particles determines the variations in load and composition of metals during build-up and wash-off processes, resulting in changes to stormwater quality at a given point in time. Reliable information about such changes is critical for designing effective stormwater pollution mitigation strategies, and that information can be generated by quantifying the uncertainty inherent to build-up and wash-off processes. It is noteworthy that although this chapter discussed metals, other pollutants attached to particles such as hydrocarbons could also undergo similar processes and consequent variability.

References

Gbeddy, G., Jayarathne, A., Goonetilleke, A., Ayoko, G. A., & Egodawatta, P. (2018). Variability and uncertainty of particle build-up on urban road surfaces. *Science of the Total Environment, 640–641,* 1432–1437. https://doi.org/10.1016/j.scitotenv.2018.05.384.

MikeUrban. (2014). *Mouse pollution transport—Reference manual.* Danish Hydraulic Institue.

Wijesiri, B., Egodawatta, P., McGree, J., & Goonetilleke, A. (2015). Incorporating process variability into stormwater quality modelling. *Science of the Total Environment, 533,* 454–461. https://doi.org/10.1016/j.scitotenv.2015.07.008.

Wijesiri, B., Egodawatta, P., McGree, J., & Goonetilleke, A. (2016). Influence of uncertainty inherent to heavy metal build-up and wash-off on stormwater quality. *Water Research, 91,* 264–276.

Wijesiri, B., Liu, A., Gunawardana, C., Hong, N., Zhu, P., Guan, Y., et al. (2018). Influence of urbanisation characteristics on the variability of particle-bound metals build-up: A comparative study between China and Australia. *Environmental Pollution, 242,* 1067–1077. https://doi.org/10.1016/j.envpol.2018.07.123.

Chapter 5
Practical Implications and Recommendations for Future Research

Abstract This chapter discusses implications of the outcomes of this research study in relation to stormwater pollution mitigation. It also presents a novel approach for implementing the research outcomes, using the concept of Logic Models. This approach is a logical sequence to a set of activities that enable the designing of effective stormwater pollution mitigation strategies utilising reliable information generated by stormwater quality models. The chapter further identifies opportunities for future research, namely, pollutant-particulate interactions, uncertainty assessment in relation to models with different complexity, and the need for integrating different modelling frameworks to improve the quantification of overall uncertainty associated with stormwater quality modelling outcomes.

Keywords Logic models · Modelling uncertainty · Pollutant processes Process uncertainty · Stormwater quality modelling

5.1 Background

Currently, urban water managers undertake decision making in relation to stormwater pollution mitigation using stormwater quality predictions without a complete understanding of the uncertainty arising from a number of sources. This can result in the risk of failure of strategies to improve stormwater quality, posing significant impact on both, human and aquatic ecosystem health. In the previous chapters, the need for quantitatively assessing the uncertainty associated with stormwater pollutant processes to enhance stormwater pollution mitigation was highlighted. Further, stormwater management personnel may lack adequate understanding on how to accurately interpret the outcomes of uncertainty assessment. This is due to little or no research on uncertainty inherent to stormwater pollutant processes and the lack of in-built tools for uncertainty assessment in current stormwater quality modelling software.

Accordingly, this chapter initially evaluates a number of decisions that can be made at different levels of uncertainty, and the risk associated with those decisions.

© The Author(s), under exclusive license to Springer Nature Singapore Pte Ltd. 2019
B. Wijesiri et al., *Decision Making with Uncertainty in Stormwater Pollutant Processes*, SpringerBriefs in Water Science and Technology,
https://doi.org/10.1007/978-981-13-3507-5_5

Then, the chapter discusses how to utilise the outcomes of the current research study to develop a practical approach to assist stormwater management personnel to understand and quantify the uncertainty associated with stormwater quality modelling outcomes, and thereby contribute to the design of effective pollution mitigation strategies.

5.2   Implications of Research Outcomes in Relation to Stormwater Pollution Mitigation

Accurate interpretation of stormwater quality modelling outcomes (predictions of stormwater quality) is the key to designing effective strategies to mitigate stormwater pollution in urban areas. However, stormwater quality predictions may vary over a range dictated by uncertainties arising from process variability, model input and calibration data, model parameters, and model structure. Therefore, it is important to understand how these uncertainties could influence the interpretation of model predictions and resulting planning and management decision making.

Figure 5.1 illustrates the different scenarios for decision making at different uncertainty levels. Each scenario would result in a stormwater pollution mitigation strategy that has some level of risk of failure. These decision making scenarios can be understood by considering two cases addressing the current practice and the approach proposed based on the outcomes of this research study. Accordingly, in Case 1, the decisions are made only considering modelling uncertainty, while in Case 2, decisions are made considering both, modelling and process uncertainty.

The research outcomes indicated that the models produce similar predictions (C_0) in Case 1 and Case 2, implying that Decision A would have the same consequences in both cases. On the other hand, Decision B does not account for the expected variation in C_0 (i.e. from L1 to C_0). As such, Decision B, which is made based on L1 would potentially increase the risk of failure compared to the risk of failure associated with Decision A (i.e. R1 < R2). However, Decision C, which is made based on U1, would lead to the design of mitigation strategies with reduced risk of failure compared to Decision A and Decision B (i.e. R3 < R1 < R2).

It is also evident from the research outcomes that process uncertainty has the largest influence on upper uncertainty limit of predicted stormwater quality rather than on the lower uncertainty limit. This means that Decision C and Decision E potentially would lead to pollution mitigation strategies that have significantly different performance due to the difference in the risks of failure (i.e. R5 < R3). On the other hand, the strategies designed based on Decision B and Decision D would perform similarly given that R2 ≈ R4. Intuitively, the decisions, which are made based on U1/U2 are likely to result in better performing strategies than those designed based L1/L2. This is due to the fact that the upper uncertainty limit encompasses the full range of variation of model predictions. Furthermore, Decision E (Case 2) can be identified as resulting in stormwater pollution mitigation strategies with minimal risk of failure.

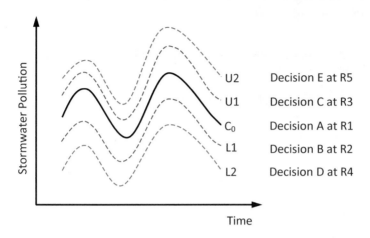

Where;

   C_0 – model prediction

   U1 – upper limit (modelling uncertainty)

   L1 – lower limit (modelling uncertainty)

   U2 – upper limit (modelling + process uncertainty)

   L2 – lower limit (modelling + process uncertainty)

   R1, R2, R3, R4 and R5 – risks of failure associated with each decision, such

                    that R5 < R3 < R1 < R2 < R4

Fig. 5.1 Influence of the uncertainty in stormwater quality modelling outcomes on planning and management decision making

Additionally, the study outcomes revealed that build-up process uncertainty has a greater influence on stormwater quality modelling outcomes compared to the wash-off process. Therefore, stormwater management personnel should ensure that planning and management decisions are made specifically accounting for the changes to pollutant load and composition during dry weather periods. Moreover, it is also evident from the current study that process uncertainty exerts different influences on stormwater quality predictions for different storm events. This further highlights the fact that the current practice in decision making, which typically considers the effects of rainfall intensity, duration and resulting runoff volume, needs to account for process uncertainty as an integral part of stormwater quality prediction.

5.2.1 Framework for Enhancing Stormwater Pollution Mitigation Strategies

Using the concept of Logic Models (McLaughlin and Jordan 1999) and the research outcomes discussed in previous chapters, this section proposes a step-by-step approach for designing effective stormwater pollution mitigation strategies. Logic Models facilitate the understanding of the process of achieving desired outcomes by following a series of interrelated activities. Logic models can help in setting the logical sequence of activities, identifying the best conditions for achieving the expected outcomes and actions necessary to optimise the performance of the approach (Carroll et al. 2006; Jordan and Mortensen 1997; Millar et al. 2001; Rush and Ogborne 1991).

5.2.2 Logic Model

A logic model replicates a specific approach with several elements, namely, resources (inputs), activities, outputs, customers, outcomes and external influential factors. Therefore, the model development requires collecting relevant information of these elements. The information required for the logic model discussed in this section can be acquired from the previous chapters. Then, understanding and clearly defining the problem of interest is key to successfully develop the approach for problem solution. The problem is defined as given below.

> The decision making in the context of stormwater pollution mitigation is affected by the lack of reliability in stormwater quality modelling outcomes. This is due to the fact that current modelling tools do not provide quantitative information on uncertainty arising from process variability and process modelling. In-depth understanding of the sources of uncertainty and robust methodologies to integrate this uncertainty into modelling outcomes and their accurate interpretation are necessary.

Figure 5.2 shows the logically organised information for each element of the Logic Model. As such, two groups of customers that benefit from the outputs of key activities can be identified. These customers are expected to create three short term/intermediate (modelling software, informed decision and treatment measures) outcomes that are critical for improving stormwater quality. The implementation of the approach is discussed in detail below.

5.2.3 Implementation of the Approach

Activity 1

This involves developing stormwater quality models that enable the quantification of process uncertainty. As such, the knowledge base created in this research is used

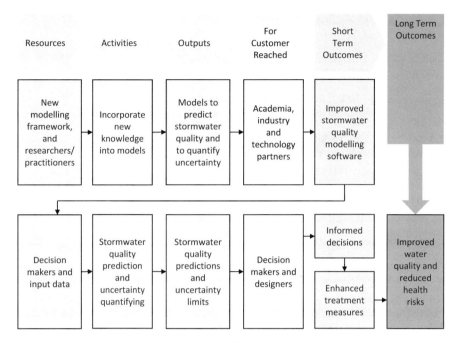

Fig. 5.2 Approach for enhancing stormwater pollution mitigation

to incorporate characteristics of the variability in pollutant processes into stormwater quality models. This can be achieved either by revising current mathematical formulations of pollutant processes, or developing new replication models incorporating key influential factors. Then, the water engineering researchers could work together with software developers to rectify the limitations in current stormwater quality modelling software, or to create new software that can quantify uncertainty as an integral part of model predictions.

Activity 2

This involves stormwater planning and management decision making based on accurate prediction and uncertainty quantification using the improved/new modelling tools. Given that the uncertainty is integrated with the model predictions, the decision makers are enabled to account for the changes in pollutant loads entrained in stormwater runoff resulting from the build-up and wash-off process variability, and in turn, will guide the designing of effective strategies to mitigate stormwater pollution. Accordingly, significant improvement in stormwater quality in urban catchments is expected in order to safeguard receiving waters and enhance urban liveability.

5.2.4 Improvements to the Approach

Understanding how external factors influence model outcomes is critical as they can exert both, positive and negative impacts on model performance.

a. *Contribution to knowledge on pollutant processes by other researchers*

The affinity of toxic stormwater pollutants to particulates changes during build-up and wash-off. This results in variations in the pollutant load and composition in stormwater runoff (see Chap. 4). These changes to pollutant-particulate relationships are primarily driven by adsorption and desorption behaviour of particles. The investigation of such complex processes is necessary in order to develop mathematical formulations in stormwater quality models. Therefore, implementation of the proposed approach requires reviewing relevant state-of-the-art knowledge.

b. *Modelling uncertainty*

The approach proposed in the Logic Model specifically addresses the quantification of uncertainty inherent to pollutant processes. However, the interpretation of stormwater quality modelling outcomes should also consider modelling uncertainty arising from model structure, model parameters, and input data and calibration data. Given that current techniques available for quantitatively assessing modelling uncertainty has limitations (see Chap. 1), the proposed approach needs to be implemented together with new developments in uncertainty assessment techniques to design better performing stormwater pollution mitigation strategies.

5.3 Recommendations for Future Research

This manuscript primarily contributes to the understanding of the intrinsic variability and resulting uncertainty in pollutant build-up and wash-off processes in urban catchments. Additionally, it also identifies future research opportunities that can contribute to the design of effective stormwater pollution mitigation strategies.

a. *Pollutant-particulate interactions*

While particle behaviour induces process variability, it is also implied that the interactions between particles and other pollutants underpinned by adsorption and desorption processes also contribute to variations in pollutant load and composition during dry weather periods and storm events. Due to inadequate knowledge of these interactions, stormwater quality modelling tools lack accurate mathematical descriptions of the processes that particle-bound pollutants undergo.

b. *Uncertainty assessment in relation to complex models*

The methodology for assessing uncertainty discussed in this manuscript has been developed using the simplest mathematical replications of build-up and wash-off processes. As more comprehensive replication models are necessary for incorporating the key influential factors in relation to these processes (such as vehicular traffic, land use and effects of pollutant re-distribution), the current methodology can be further investigated for its performance with varying complexity using different models.

c. *Use of statistical frameworks to improve the quantification of modelling uncertainty*

The current methodology for assessing uncertainty focuses on accounting for inherent process uncertainty. However, accounting for modelling uncertainty, which arises from sources such as model structure and input data, is critical for informed decision making in relation to stormwater pollution mitigation. As such, the methodology proposed in this manuscript can be further developed by combining with different modelling frameworks. In this regard, integration with Bayesian statistical techniques would be of significant benefit due to their inherent capability to account for several uncertainty sources, in particular, input data and model parameters, facilitating the quantification of overall uncertainty associated with stormwater quality predictions.

References

Carroll, S., Goonetilleke, A., Thomas, E., Hargreaves, M., Frost, R., & Dawes, L. (2006). Integrated risk framework for onsite wastewater treatment systems. *Environmental Management, 38*(2), 286–303. https://doi.org/10.1007/s00267-005-0280-5.

Jordan, G., & Mortensen, J. (1997). Measuring the performance of research and technology programs: A balanced scorecard approach. *The Journal of Technology Transfer, 22*(2), 13–20. https://doi.org/10.1007/BF02509640.

McLaughlin, J. A., & Jordan, G. B. (1999). Logic models: A tool for telling your programs performance story. *Evaluation and Program Planning, 22*(1), 65–72. https://doi.org/10.1016/S0149-7189(98)00042-1.

Millar, A., Simeone, R. S., & Carnevale, J. T. (2001). Logic models: A systems tool for performance management. *Evaluation and Program Planning, 24*(1), 73–81. https://doi.org/10.1016/S0149-7189(00)00048-3.

Rush, B., & Ogborne, A. (1991). Program logic models: Expanding their role and structure for program planning and evaluation. *The Canadian Journal of Program Evaluation, 6*(2), 95.

Appendix A

See Figs. A.1, A.2, A.3, A.4, A.5 and A.6.

Fig. A.1 Aerial view of the study catchment and street views of road sites relating to historical data

B. Wijesiri et al., *Decision Making with Uncertainty in Stormwater Pollutant Processes*, SpringerBriefs in Water Science and Technology, https://doi.org/10.1007/978-981-13-3507-5

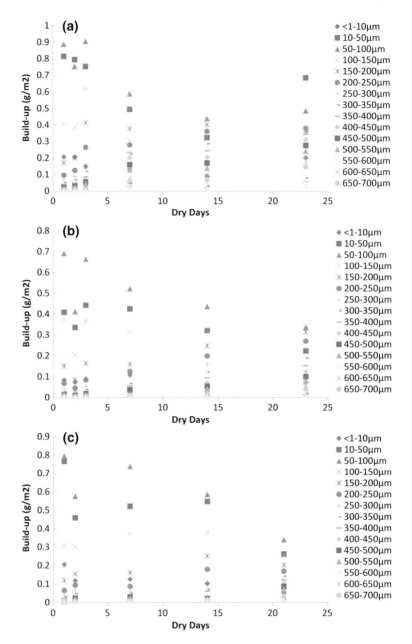

Fig. A.2 Temporal variation in the build-up of particulate solids: **a** Gumbeel Court; **b** Lauder Court; **c** Piccadilly Place (adapted from Wijesiri et al. 2015c)

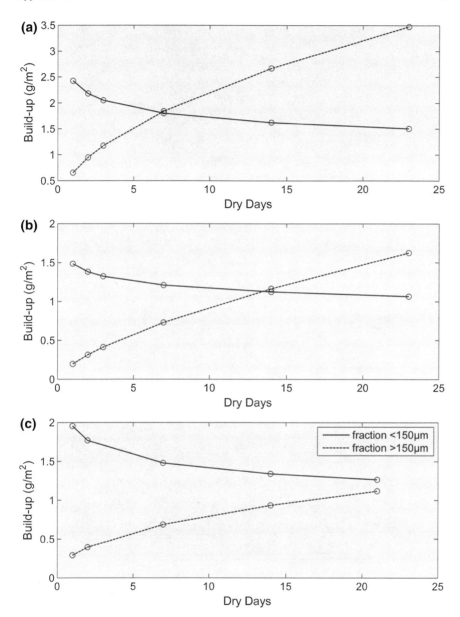

Fig. A.3 Predicted build-up patterns for particles <150μm and >150μm: **a** Gumbeel Court; **b** Lauder Court; **c** Piccadilly Place (adapted from Wijesiri et al. 2015c)

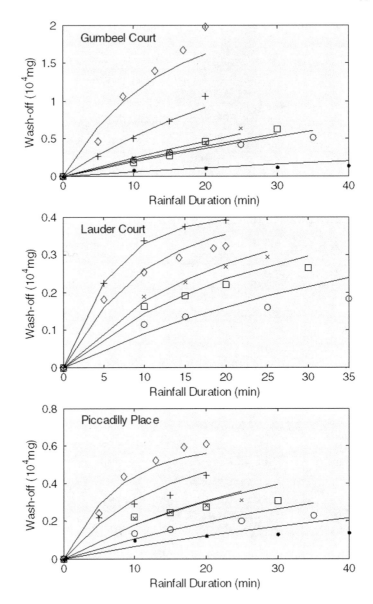

Fig. A.4 Predicted wash-off patterns for particles <150 μm

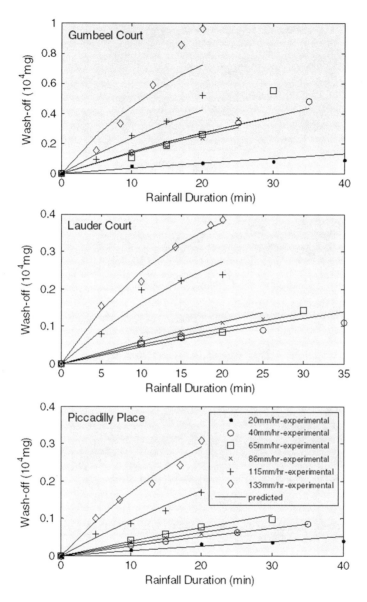

Fig. A.5 Predicted wash-off patterns for particles >150μm

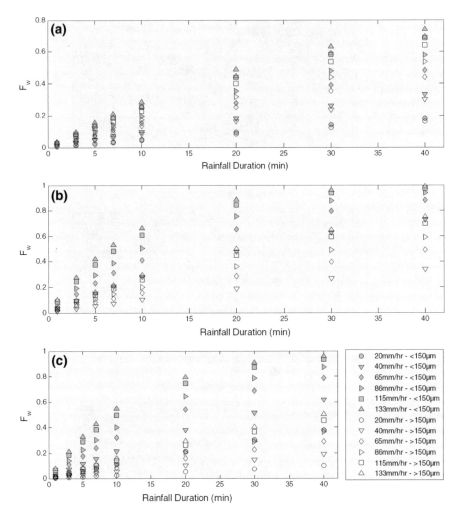

Fig. A.6 Predicted patterns of fraction wash-off (F_w): **a** Gumbeel Court; **b** Lauder Court; **c** Piccadilly Place (adapted from Wijesiri et al. 2015b)

Appendix B

See B.1, B.2, B.3, B.4, B.5, B.6, B.7, B.8, B.9 and Tables B.1, B.2.
   See Figs. B.10, B.11, B.12, B.13 and B.14.

B. Wijesiri et al., *Decision Making with Uncertainty in Stormwater Pollutant Processes*, SpringerBriefs in Water Science and Technology, https://doi.org/10.1007/978-981-13-3507-5

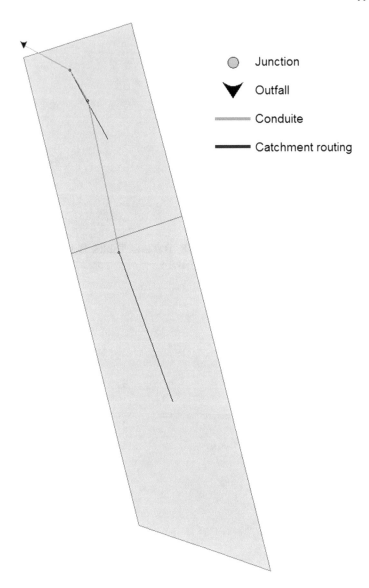

Fig. B.1 Gumbeel Court catchment model (adapted from Wijesiri et al. 2016)

Fig. B.2 Highland Park catchment model (adapted from Wijesiri et al. 2016)

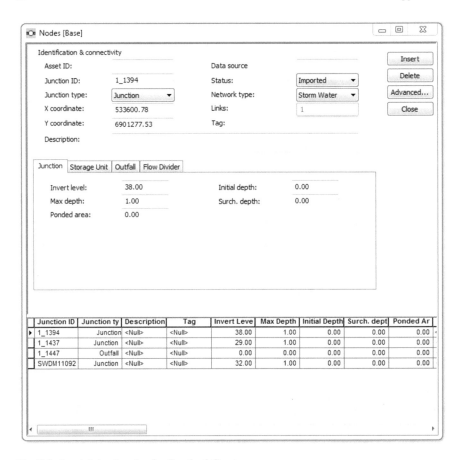

Fig. B.3 Input data of nodes for Gumbeel Court

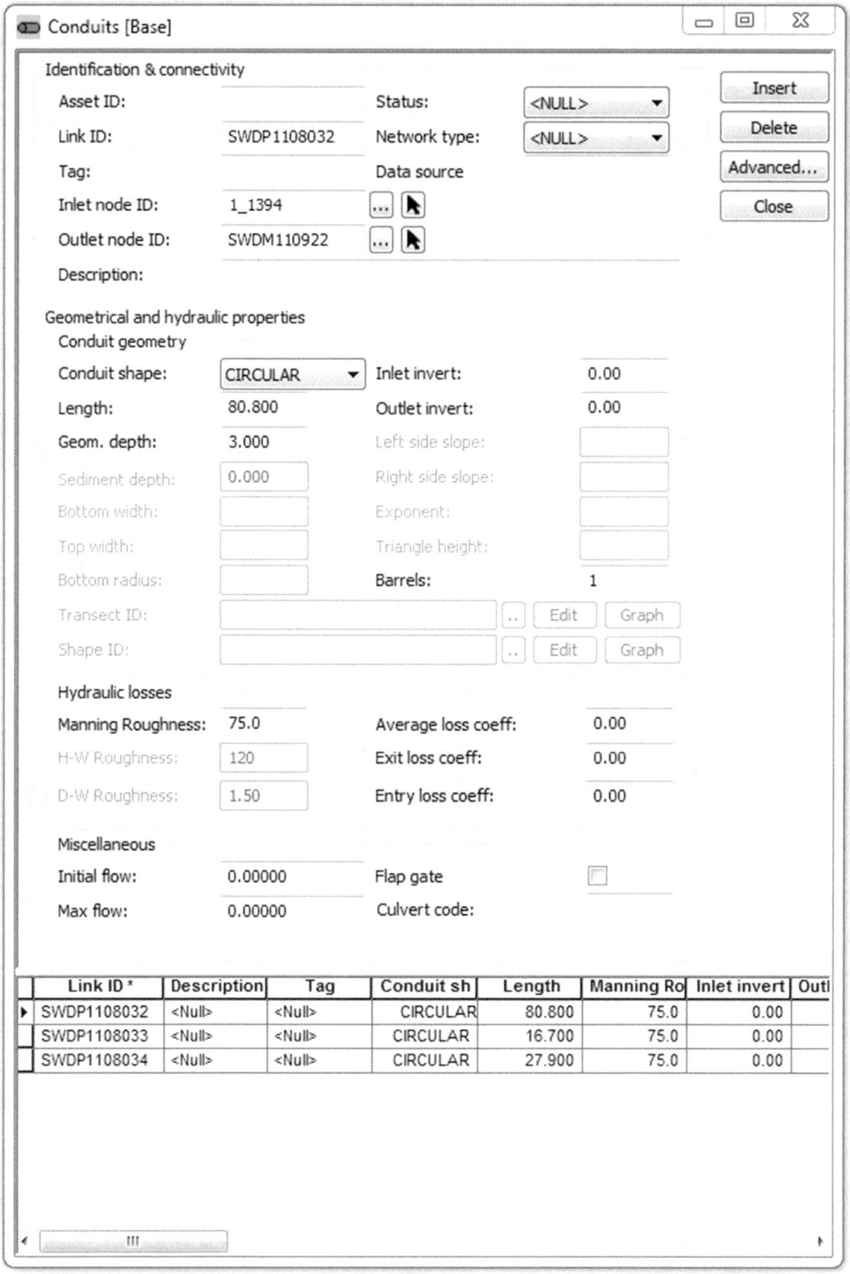

Fig. B.4 Input data of conduits for Gumbeel Court

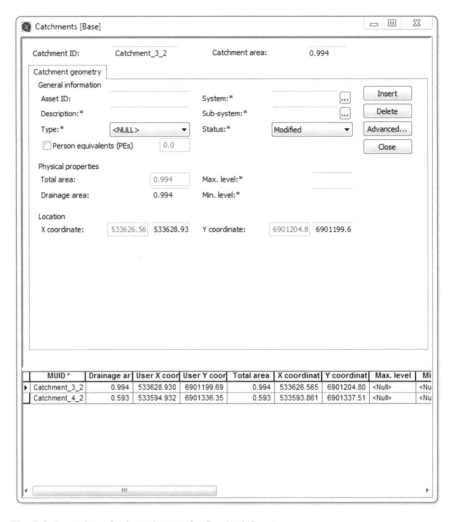

Fig. B.5 Input data of sub-catchments for Gumbeel Court

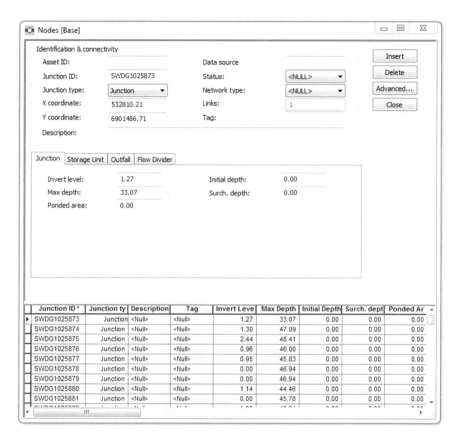

Fig. B.6 Input data of nodes for Highland Park

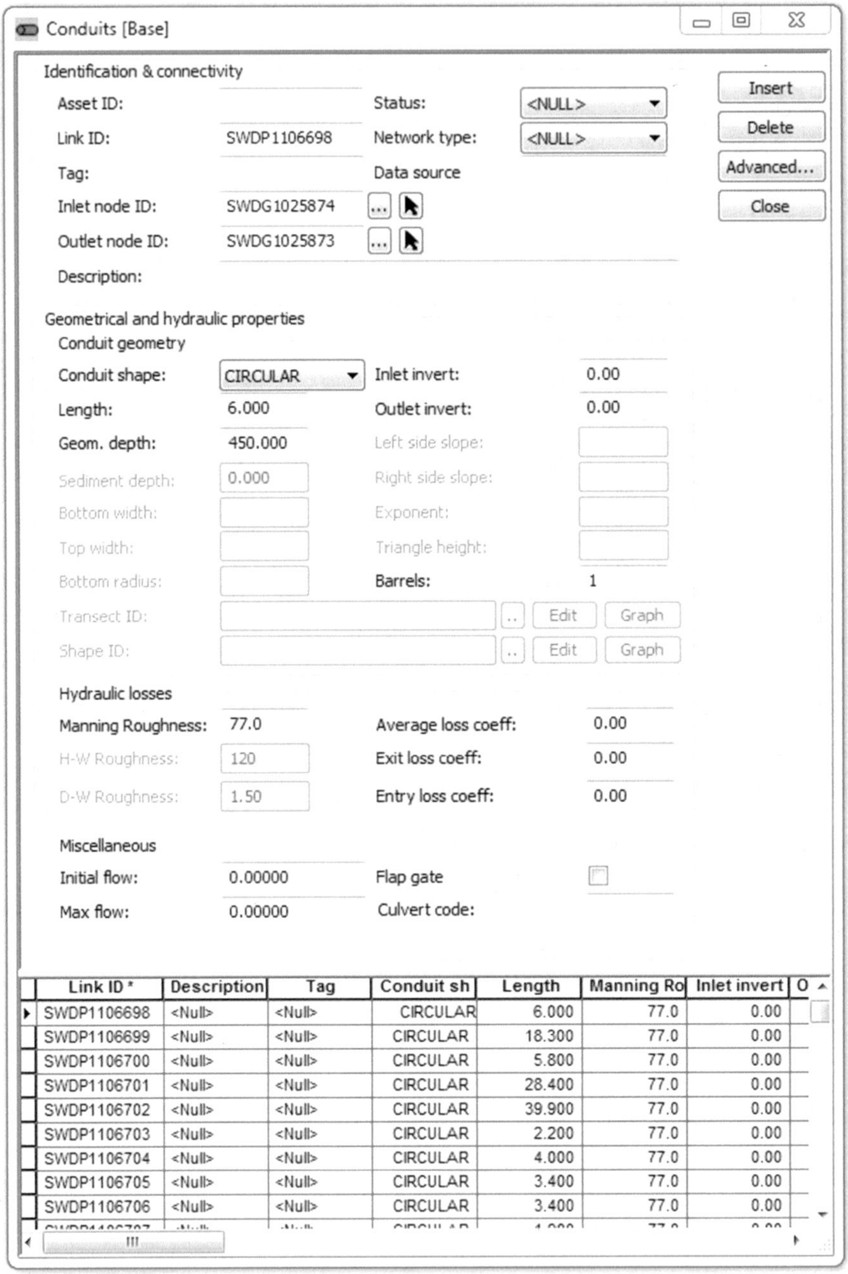

Fig. B.7 Input data of conduits for Highland Park

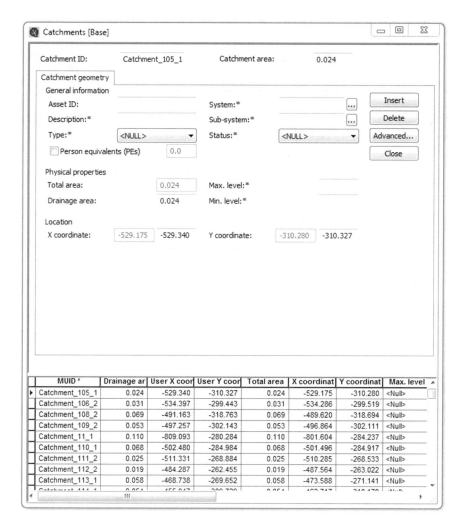

Fig. B.8 Input data of sub-catchments for Highland Park

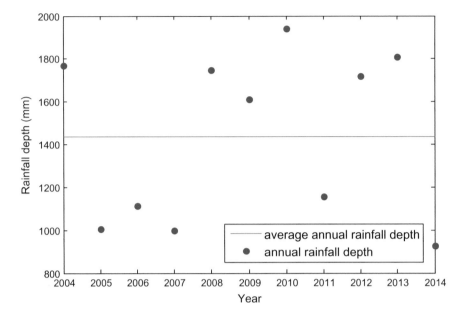

Fig. B.9 Variation of annual rainfall depth from 2004 to 2014 (adapted from Wijesiri et al. 2016)

Table B.1 Characteristics of storm events selected for stormwater quality modelling and simulated runoff (adapted from Wijesiri et al. 2016)

Event no.	Date	Average intensity (mm/h)	Duration (h)	Runoff (m³)	
				Gumbeel Court	Highland Park
1	17/1/2005	23.0	0.58	23.38	1583.16
2	23/1/2005	31.2	0.17	3.62	230.53
3	24/1/2005	44.0	0.25	18.46	1221.05
4	25/1/2005	21.8	0.18	1.30	85.26
5	2/2/2005	42.8	0.27	18.99	1260.00
6	27/4/2005	25.1	0.18	2.44	156.84
7	28/4/2005	24.7	0.28	7.96	512.63
8	15/6/2005	22.3	0.23	3.86	245.26
9	28/6/2005	32.5	0.28	12.96	847.37
10	29/6/2005	35.1	1.10	82.67	5775.79
11	30/6/2005	23.0	4.33	228.06	16,109.47
12	11/9/2005	22.6	0.28	6.65	425.26
13	16/9/2005	32.3	0.22	7.74	494.74
14	21/10/2005	27.3	0.18	3.28	209.47
15	24/10/2005	50.5	0.72	160.75	10,391.58
16	2/11/2005	26.9	0.35	13.63	898.95

(continued)

Table B.1 (continued)

Event no.	Date	Average intensity (mm/h)	Duration (h)	Runoff (m³) Gumbeel Court	Highland Park
17	4/11/2005	32.7	0.30	14.39	945.26
18	5/11/2005	48.0	0.15	8.53	551.58
19	6/11/2005	26.4	0.17	2.01	130.53
20	13/11/2005	26.8	0.22	5.11	325.26
21	24/11/2005	35.3	0.53	35.69	2444.21
22	2/12/2005	25.1	1.43	76.75	5372.63
23	3/12/2005	23.4	0.32	9.01	583.16
24	8/12/2005	33.1	0.70	46.09	3185.26
25	13/12/2005	34.6	0.28	14.29	936.84
26	16/12/2005	57.2	0.98	337.16	21,635.79
27	25/12/2005	26.0	0.50	22.16	1500.00
1	3/1/2008	13.8	0.22	0.03	3.16
2	4/1/2008	26.3	3.25	194.23	13,704.21
3	5/1/2008	25.1	0.18	2.44	156.84
4	15/1/2008	25.5	0.13	0.36	25.26
5	30/1/2008	21.8	0.18	1.30	85.26
6	2/2/2008	23.0	0.20	2.50	160.00
7	3/2/2008	64.7	2.20	1101.11	72,353.68
8	4/2/2008	18.3	0.32	5.38	344.21
9	6/2/2008	74.2	0.18	49.26	3222.11
10	26/2/2008	21.6	0.42	12.84	849.47
11	26/3/2008	66.5	0.22	50.87	3328.42
12	27/3/2008	19.7	0.47	13.38	889.47
13	28/3/2008	19.8	0.77	27.57	1894.74
14	6/4/2008	18.0	0.17	0.05	3.16
15	28/5/2008	25.4	0.28	8.40	542.11
16	30/5/2008	17.5	0.18	0.17	12.63
17	1/6/2008	19.0	0.20	1.00	65.26
18	2/6/2008	15.9	1.03	30.72	2125.26
19	3/6/2008	33.3	0.37	20.05	1341.05
20	18/6/2008	19.2	0.25	3.04	193.68
21	20/6/2008	23.1	0.23	4.26	271.58
22	5/9/2008	54.9	0.23	33.79	2235.79
23	11/10/2008	22.7	0.30	7.60	489.47
24	21/10/2008	32.4	0.17	4.04	257.89
25	8/11/2008	29.8	0.45	23.00	1553.68
26	13/11/2008	18.3	0.67	20.58	1401.05
27	16/11/2008	22.3	0.35	9.97	650.53

(continued)

Table B.1 (continued)

Event no.	Date	Average intensity (mm/h)	Duration (h)	Runoff (m^3)	
				Gumbeel Court	Highland Park
28	19/11/2008	17.5	0.87	27.65	1903.16
29	20/11/2008	19.2	0.87	31.07	2143.16
30	26/11/2008	30.8	1.18	77.67	5429.47
31	29/11/2008	46.5	0.27	24.99	1666.32
32	6/12/2008	55.3	0.30	54.58	3588.42
33	7/12/2008	18.0	0.20	0.68	45.26
34	11/12/2008	25.3	0.15	0.90	61.05
35	19/12/2008	22.2	0.22	2.98	189.47
36	24/12/2008	45.0	0.33	33.05	2221.05
37	28/12/2008	28.2	0.85	48.13	3337.89
38	29/12/2008	25.2	0.33	11.27	736.84

Table B.2 Estimated build-up and wash-off coefficients (adapted from Wijesiri et al. 2016)

Catchment	Process	Coefficient	Classical model	Revised model	
				Fraction <150 μm	Fraction >150 μm
Gumbeel Court	Build-up	a	2.77	2.43	0.544
		b	0.173	0.154	0.618
	Wash-off	k	0.0109	0.0103	0.0119
Highland Park	Build-up	a	2.22	1.97	0.399
		b	0.113	0.143	0.528
	Wash-off	k	0.0156	0.0238	0.00920

Note proportional error model was specified in all estimations

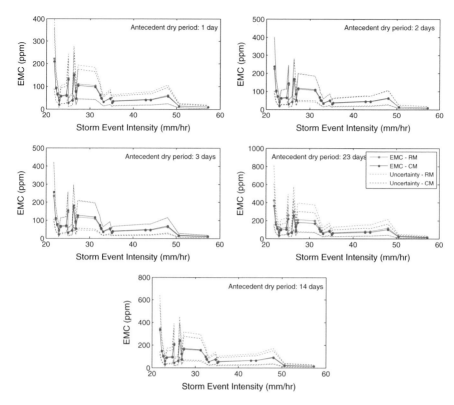

Fig. B.10 Predicted Event Mean Concentration (EMC) of particulate solids and associated uncertainty—Gumbeel Court, storm events in 2005. *Note* CM—classical models, and RM—revised models

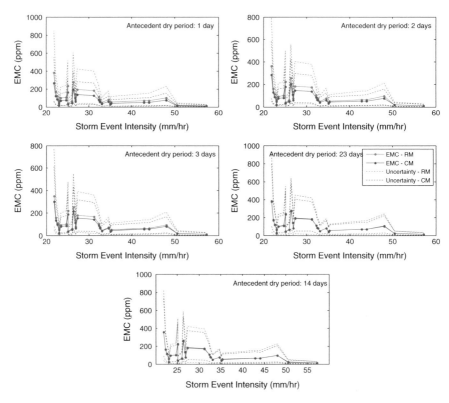

Fig. B.11 Predicted Event Mean Concentration (EMC) of particulate solids and associated uncertainty—Highland Park, storm events in 2005. *Note* CM—classical models, and RM—revised models

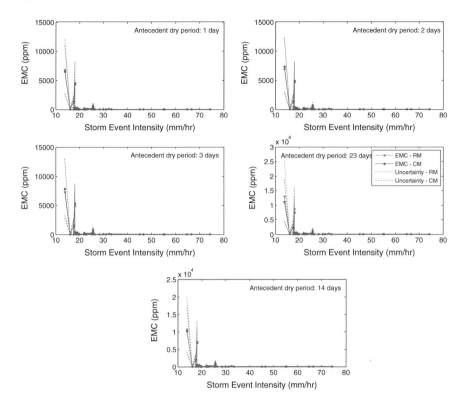

Fig. B.12 Predicted Event Mean Concentration (EMC) of particulate solids and associated uncertainty—Gumbeel Court, storm events in 2008. *Note* CM—classical models, and RM—revised models

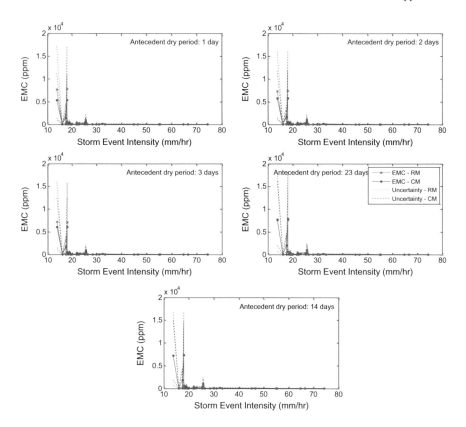

Fig. B.13 Predicted Event Mean Concentration (EMC) of particulate solids and associated uncertainty—Highland Park, storm events in 2008. *Note* CM—classical models, and RM—revised models

Gumbeel Court

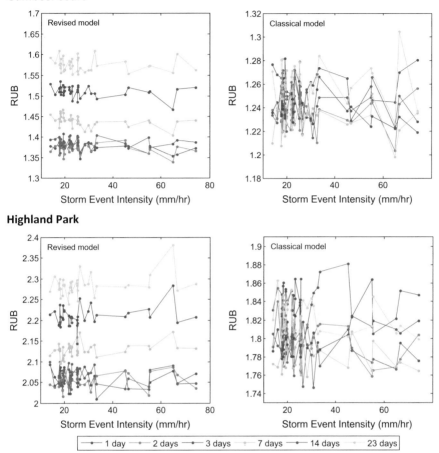

Fig. B.14 Relative uncertainty bandwidth (RUB) for classical and revised models—2008

Index

Printed in the United States
By Bookmasters